DESCRIPTIVE
CHEMISTRY

DONALD A. McQUARRIE

PETER A. ROCK

Department of Chemistry
University of California at Davis

W. H. FREEMAN AND COMPANY ▪ NEW YOR

The cover shows a platinum metal crucible heated by a Meker burner. Platinum crucibles are used as reaction vessels in chemical analysis because of their remarkable chemical inertness even at high temperatures.

(Photo courtesy of The Engelhard Corporation.)

Library of Congress Cataloging in Publication Data
McQuarrie, Donald A. (Donald Allan)
Descriptive chemistry.
Includes index.
1. Chemistry. I. Rock, Peter A., 1939–
II. Title.
QD33.M246 1984 546 84-21078
ISBN 0-7167-1706-9 (pbk.)

Printed in the United States of America

1 2 3 4 5 6 7 8 9 0 KP 3 2 1 0 8 9 8 7 6 5

CONTENTS

PREFACE

We have written this short text on descriptive chemistry for the many general chemistry courses that cover descriptive chemistry in the latter part of the course. A knowledge of the chemistry of the elements is essential for an understanding of the many remarkable and fascinating aspects of chemical behavior, and we present a systematic discussion of the chemistry of the elements and their compounds, with emphasis on the correlation between chemical properties and the periodic table. Chapter 1 consists of a discussion of periodic trends in both physical and chemical properties. Vertical and horizontal trends in metallic versus nonmetallic character, atomic and ionic size, ionization energy, electronegativity, acidity of oxides, oxidation states, relative stabilities of hydrides, and the differences between the first member of a group and the subsequent group members are discussed. Following Chapter 1, we discuss hydrogen (Chapter 2), the alkali metals (Chapter 3), the alkaline earth metals (Chapter 4), the Group 3 through the Group 6 elements (Chapters 5 through 8), the halogens (Chapter 9), the noble gases (Chapter 10), and the transition metals (Chapter 11). In Chapter 11, we concentrate on the 3*d* transition metals and gold, silver, and mercury, discussing the metallurgy of both iron and copper in some detail.

Although we initially planned this text to supplement our *General Chemistry* (W. H. Freeman and Company, New York, 1984), we have written it to be used independently. The level is consistent with any of a number of general chemistry texts, and could be used in conjunction with any of them. In the interest of brevity, we have not included material on coordination chemistry, nuclear chemistry, or organic chemistry, as these topics are discussed systematically in our *General Chemistry*, as well as in most other general chemistry books.

We have included about 15 to 25 questions at the end of each of the 11 chapters. Most of the questions require qualitative answers, but some of them involve stoichiometric, thermodynamic, and electrochemical calculations. Answers to all the questions are given in Appendix B.

We wish to thank a number of people for their contribution to this text: Carole McQuarrie, for reading the entire manuscript and for checking the answers to all the questions; Travis Amos, for researching most of the photographs and for many excellent suggestions; Chip Clark, for setting up and taking many of the photographs of chemicals and chemical reactions; Neil Patterson and Linda Chaput, for their constant support and integrity; Lee Walters, for her participation in the initial stages of this project; Betsy Galbraith, James Maurer, and Karen McDermott for their outstanding contributions to the production of this book; and Elaine Rock, for typing the manuscript in her usual excellent manner.

Donald A. McQuarrie
Peter A. Rock

August 1984

DESCRIPTIVE CHEMISTRY

PERIODICITY AND THE PERIODIC TABLE

1	2											3	4	5	6	7	8	Period	Subshells
1 **H**																	2 **He**	1	
3 **Li**	4 **Be**											5 **B**	6 **C**	7 **N**	8 **O**	9 **F**	10 **Ne**	2	2*s*, 2*p*
11 **Na**	12 **Mg**											13 **Al**	14 **Si**	15 **P**	16 **S**	17 **Cl**	18 **Ar**	3	3*s*, 3*p*
19 **K**	20 **Ca**	21 **Sc**	22 **Ti**	23 **V**	24 **Cr**	25 **Mn**	26 **Fe**	27 **Co**	28 **Ni**	29 **Cu**	30 **Zn**	31 **Ga**	32 **Ge**	33 **As**	34 **Se**	35 **Br**	36 **Kr**	4	4*s*, 3*d*, 4*p*
37 **Rb**	38 **Sr**	39 **Y**	40 **Zr**	41 **Nb**	42 **Mo**	43 **Tc**	44 **Ru**	45 **Rh**	46 **Pd**	47 **Ag**	48 **Cd**	49 **In**	50 **Sn**	51 **Sb**	52 **Te**	53 **I**	54 **Xe**	5	5*s*, 4*d*, 5*p*
55 **Cs**	56 **Ba**	71 **Lu**	72 **Hf**	73 **Ta**	74 **W**	75 **Re**	76 **Os**	77 **Ir**	78 **Pt**	79 **Au**	80 **Hg**	81 **Tl**	82 **Pb**	83 **Bi**	84 **Po**	85 **At**	86 **Rn**	6	6*s*, 5*d*, 6*p*
87 **Fr**	88 **Ra**	103 **Lr**	104 **Unq**	105 **Unp**	106 **Unh**	107 **Uns**	108	109 **Une**										7	7*s*, 6*d*

Lanthanide series

57 **La**	58 **Ce**	59 **Pr**	60 **Nd**	61 **Pm**	62 **Sm**	63 **Eu**	64 **Gd**	65 **Tb**	66 **Dy**	67 **Ho**	68 **Er**	69 **Tm**	70 **Yb**	6	4*f*

Actinide series

89 **Ac**	90 **Th**	91 **Pa**	92 **U**	93 **Np**	94 **Pu**	95 **Am**	96 **Cm**	97 **Bk**	98 **Cf**	99 **Es**	100 **Fm**	101 **Md**	102 **No**	7	5*f*

A periodic table indicating which orbitals are occupied by the valence electrons of each element. Blocks of elements with the same shading have the same valence electron subshells.

The 108 known elements are classified as either *metals, nonmetals,* or *semimetals.* As is shown on the inside front cover, 83 of the elements are classified as metals, 17 are classified as nonmetals, and 8 are classified as semimetals.

At 25°C, all the metallic elements except mercury are solids. Pure metals, as opposed to nonmetals and semimetals, have a lustrous appearance, and readily conduct heat and electricity. Metals can be squeezed or hammered into various shapes *(malleable)* and drawn into wires *(ductile).* The oxides of metals are basic (for example, Na_2O), and the oxides, hydrides, and halides of metals are often ionic compounds with high melting points (Table 1-1).

Table 1-1 The melting points and boiling points of the chlorides of the second-row and third-row elements

Second-row elements

compound	LiCl	$BeCl_2$	BCl_3	CCl_4	NCl_3	OCl_2	FCl
melting point/°C	614	405	−107	−23	−45	−20	−154
boiling point/°C	1350	520	13	77	65	3.8	−101

Third-row elements

compound	NaCl	$MgCl_2$	$AlCl_3$	$SiCl_4$	PCl_3	SCl_2	Cl_2
melting point/°C	800	708	170	−70	−112	−78	−101
boiling point/°C	1413	1412	183	58	76	60	−35

The nonmetals vary greatly in appearance. At 25°C and 1 atm, carbon, phosphorus, sulfur, selenium, and iodine are solids; bromine is a liquid; hydrogen, oxygen, nitrogen, fluorine, and chlorine are diatomic gases; and helium, neon, argon, krypton, xenon, and radon are monatomic gases. The nonmetals are insulators (very low thermal and electrical conductivities) and are neither malleable nor ductile. The oxides of nonmetals are acidic (for example, SO_2), and the oxides, hydrides, and halides (for example, NO_2, H_2S, CCl_4) are volatile covalent compounds with low melting points (Table 1-1).

The semimetals have properties intermediate between those of the metals and the nonmetals. For example, the semimetals are called *semiconductors* because of their intermediate electrical and thermal conductivities. The semimetals are brittle.

1-1 THE METALLIC CHARACTER OF THE ELEMENTS INCREASES MOVING DOWN A COLUMN AND MOVING RIGHT TO LEFT ACROSS A ROW

The contrasting properties of the metals, the semimetals, and the nonmetals are given in Table 1-2. The metallic character of the elements increases as we move down a column in the periodic table, and as we move from right to left in a row of the periodic table (Figure 1-1). Thus francium is the most metallic element and fluorine is the most nonmetallic element. The noble gases (Group 8) are nonmetals, but they are so unreactive that, with the exception of xenon, they have very limited chemistries. Consequently, the noble gases are discussed separately. The transition from metallic to semimetallic to nonmetallic behavior as we move from left to right across the periodic table is not sharp; rather, there is a gradual change in properties from distinctly metallic (Group 1) to distinctly nonmetallic (Group 7).

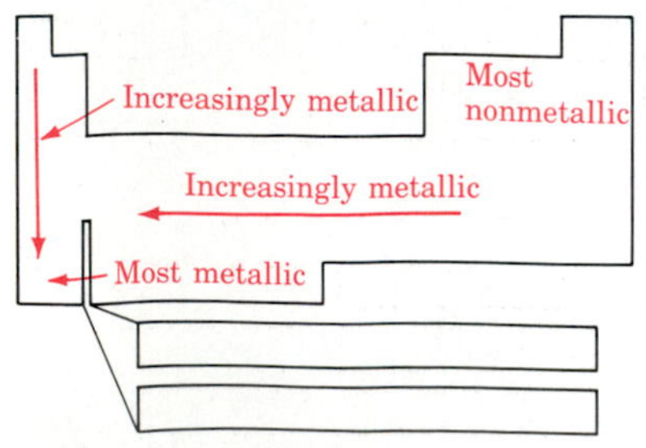

Figure 1-1 Trends in the metallic character of the elements.

Table 1-2 Comparison of physical properties of metals, semimetals, and nonmetals

Metals	*Semimetals*	*Nonmetals*
$0.7 \leq$ electronegativity < 1.8	$1.8 \leq$ electronegativity < 2.2	$2.2 \leq$ electronegativity < 4.0
basic oxides	amphoteric or weakly acidic oxides	acidic oxides
high electrical and thermal conductance	intermediate electrical and thermal conductance	insulators
electrical resistance increases with increasing temperature	electrical resistance decreases with increasing temperature	resistance insensitive to temperature
malleable and ductile	brittle	not malleable, not ductile
nonvolatile and high-melting oxides, halides, and hydrides	volatile and low-melting halides and hydrides	volatile and low-melting oxides, halides, and hydrides

1-2 OXIDES BECOME MORE BASIC MOVING DOWN A COLUMN AND MOVING RIGHT TO LEFT ACROSS A ROW

Because the metallic character of the elements in a given group increases as we go down the column, the oxides become more basic. This effect can be seen in the Group 5 elements. The oxides of nitrogen and phosphorus are acidic:

$$N_2O_3(s) + 2NaOH(aq) \rightarrow 2NaNO_2(aq) + H_2O(l)$$
$$P_4O_6(s) + 8NaOH(aq) \rightarrow 4Na_2HPO_3(aq) + 2H_2O(l)$$

The oxides of arsenic and antimony are amphoteric; for example,

$$Sb_2O_3(s) + 6HCl(aq) \rightarrow 2SbCl_3(aq) + 3H_2O(l)$$
$$Sb_2O_3(s) + 6NaOH(aq) \rightarrow 2Na_3SbO_3(aq) + 3H_2O(l)$$

and bismuth oxide is basic, being insoluble in NaOH(aq) but soluble in HCl(aq):

$$Bi_2O_3(s) + 6HCl(aq) \rightarrow 2BiCl_3(aq) + 3H_2O(l)$$

Similarly, because the metallic character of the main group elements decreases from left to right across a given row or period in the periodic table, the acidity of the oxides should increase from left to right (Figure 1-2). Taking the third row as an example, we have

basic oxides
$$\begin{cases} Na_2O(s) + 2HCl(aq) \rightarrow 2NaCl(aq) + H_2O(l) \\ MgO(s) + 2HCl(aq) \rightarrow MgCl_2(aq) + H_2O(l) \end{cases}$$

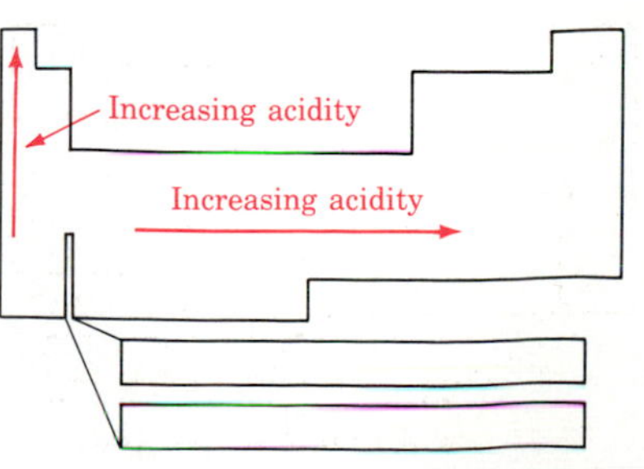

Figure 1-2 Trends in the acidity of the oxides of the elements.

amphoteric oxide

$$\begin{cases} Al_2O_3(s) + 6HCl(aq) \rightarrow 2AlCl_3(aq) + 3H_2O(l) \\ Al_2O_3(s) + 2NaOH(aq) + 3H_2O(l) \rightarrow 2NaAl(OH)_4(aq) \end{cases}$$

acidic oxides

$$\begin{cases} SiO_2(s) + 4NaOH(aq) \rightarrow Na_4SiO_4(aq) + 2H_2O(l) \\ P_2O_5(s) + 6NaOH(aq) \rightarrow 2Na_3PO_4(aq) + 3H_2O(l) \\ SO_3(g) + 2NaOH(aq) \rightarrow Na_2SO_4(aq) + H_2O(l) \\ Cl_2O_7(s) + 2NaOH(aq) \rightarrow 2NaClO_4(aq) + H_2O(l) \end{cases}$$

1-3 ELEMENTS ARE CLASSIFIED AS MAIN-GROUP ELEMENTS OR TRANSITION METALS

The theoretical basis of the periodic table in terms of the ground-state electron configurations of atoms is discussed in Chapter 8 of the text, and Figure 8-6 shows the outer electron configurations of the atoms arranged in a periodic table. Because the elements in a given group have similar outer-shell electron configurations, the simple compounds of the elements in a given group usually will have similar chemical formulas. This can be seen in Table 1-3, which shows the known binary hydrides of the main-group elements.

Table 1-3 The binary hydrides of the main-group elements

LiH	BeH_2	BH_3	CH_4	NH_3	H_2O	HF
NaH	MgH_2	AlH_3	SiH_4	PH_3	H_2S	HCl
KH	CaH_2	GaH_3	GeH_4	AsH_3	H_2Se	HBr
RbH	SrH_2	InH_3	SnH_4	SbH_3	H_2Te	HI
CsH	BaH_2	TlH_3	PbH_4	BiH_3		

Recall that the elements can be classified as *s-block elements, p-block elements, d-block elements* or *f-block elements,* indicating the highest-energy orbitals that are occupied (see frontispiece). The *s*-block and *p*-block elements that appear in the same column have similar outer electron configurations. These elements are classified into Groups 1 through 8, where the numbers 1 through 8 indicate the total number of outer *s* and *p* electrons. Collectively, these elements are called the *main-group elements.* The *d*-block elements are called *transition metals* and the *f*-block elements are called *inner transition metals.* Scandium through zinc are called the 3*d* transition metals or the first transition series metals; yttrium through cadmium are called the 4*d* transition metals or the second transition series metals; lutetium through mercury are called the 5*d* transition metals or the third transition series metals. Lanthanum through ytterbium are called the *lanthanides;* and actinium through nobelium are called the *actinides.*

As we go down a particular column in the periodic table, the value of the principal quantum number, *n,* increases. In fact, the ground-state electron configurations of the *s*-block and *p*-block elements in the *n*th row involve outer-shell electrons in *ns* and *np* orbitals (see Figure 8-6 of the text). Thus, the principal quantum number labels the rows, or *periods,* in the periodic table. Because the size or extent of orbitals increases with increasing values of *n*, the sizes of atoms increase as we go down a column. Although the diffuse, cloud-like property of the electrons in atomic orbitals prevents us from assigning a fixed, unequivocal radius to an atom or ion, we can assign radii based upon models or conventions. The particular values assigned depend somewhat upon the models or conventions that are used, but useful and reliable tables of atomic and ionic radii are available. Figure 1-3 shows the relative atomic and ionic sizes of the elements arranged in the form of a periodic table. Note that the size increases in going down a group, and in going from right to left along a row in the periodic table (Figure 1-4). Note also that cations are smaller and that anions are larger than their parent atoms (see Section 9-4 of the text).

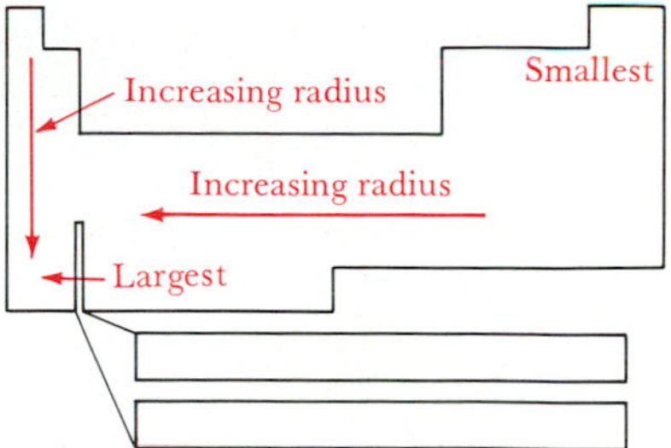

Figure 1-4 Trends in atomic radii of the elements.

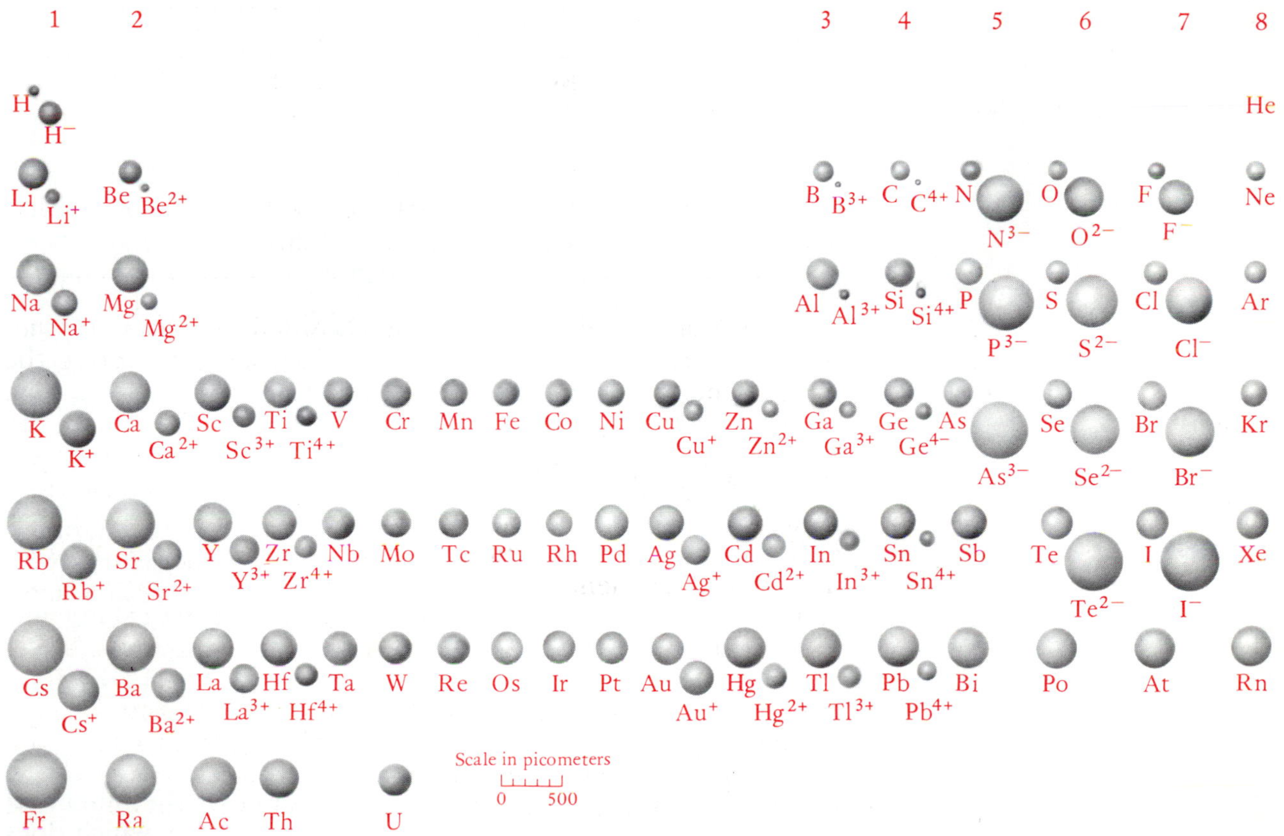

Figure 1-3 The relative sizes of atoms and ions arranged in the form of a periodic table.

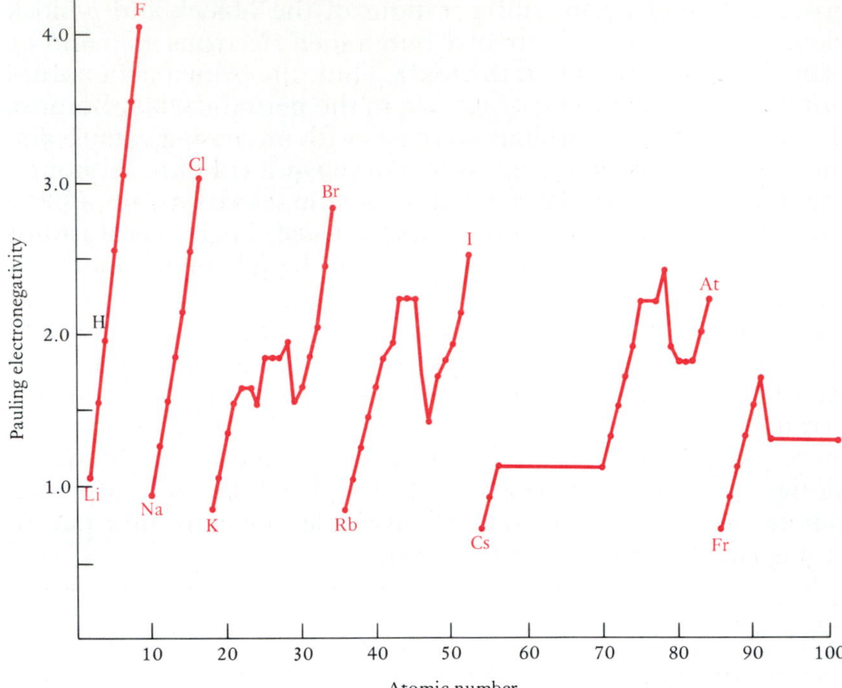

Figure 1-5 Pauling electronegativities plotted against atomic number.

1-4 ELECTRONEGATIVITY IS A PERIODIC PROPERTY

Recall that electronegativity is a measure of the force with which an atom attracts the electrons in its covalent bonds with other atoms. The larger the electronegativity, the greater the attraction of the atom for the electrons in its covalent bonds. Electronegativity is a derived quantity in that it is not directly measurable, and various electronegativity scales have been proposed. In Figure 1-5, electronegativities due to Pauling are plotted against atomic number (see also Figure 10-5 of the text). Note that electronegativities increase from left to right across the short (second and third) rows of the periodic table. This left-to-right increase in electronegativities reflects the increasingly nonmetallic nature of the elements toward the right-hand side of the table. Note also that electronegativities decrease in going down a column (Figure 1-6). The reason for the top-to-bottom decrease in electronegativities is that the nuclear attraction of the outer electrons decreases as the size of the atom increases. Note that fluorine is the most electronegative atom and that cesium and francium are the least electronegative.

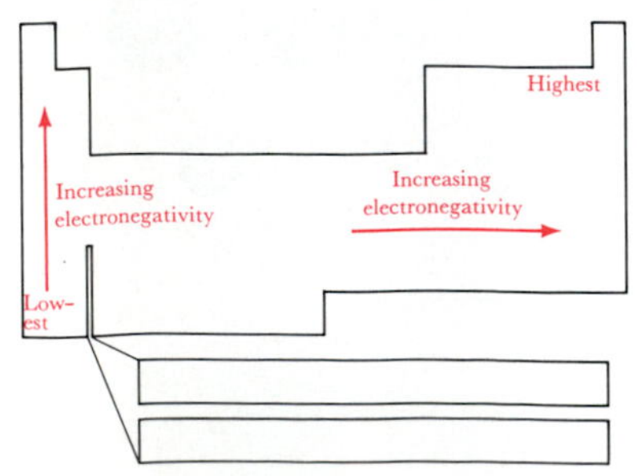

Figure 1-6 Trends in electronegativities of the elements.

1-5 MANY CHEMICAL PROPERTIES SHOW PERIODIC TRENDS

In the following chapters we shall discuss the chemical properties of the elements. We start with hydrogen, which does not fit conveniently into any group, and then go on to discuss the main group elements one group at a time, and finally we discuss the transition metals. Before doing this, however, we shall discuss some general periodic trends in chemical properties, which will be taken up in more detail in the following chapters.

The stabilities or the bond strengths of the covalent binary hydrides and halides decrease in going down a given group and generally increase in going from left to right in a given row. The bond strengths depend in a complicated way on the size and the electronegativity of the main-group element. Figure 1-7 plots the *molar bond enthalpies* of the covalent binary hydrides against group number.

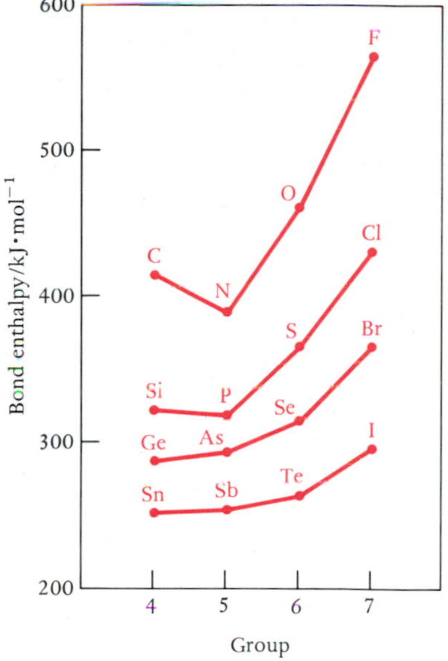

Figure 1-7 The molar bond enthalpies of the covalent binary hydrides versus group number.

The maximum oxidation state found in a main group is numerically equal to the group number. The following table of oxides illustrates this nicely.

Group	1	2	3	4	5	6	7	8
Compound	Na_2O	MgO	Ga_2O_3	SiO_2	N_2O_5	SO_3	Cl_2O_7	XeO_4
Oxidation state	+1	+2	+3	+4	+5	+6	+7	+8

Realize that not all elements in a group may occur in the maximum oxidation state. For example, fluorine, being the most electronegative element, never occurs in a positive oxidation state. Because the maximum possible oxidation state of a group is equal to the group number, it is not surprising that oxidation state is a periodic property.

Another pattern of oxidation states is that the common oxidation states of the *p*-block elements vary in steps of two from the maximum oxidation state. Using the oxyanions of nitrogen, sulfur, and chlorine as examples, we have

	$N_2O_2^{2-}$ hyponitrite (+1)	NO_2^- nitrite (+3)	NO_3^- nitrate (+5)
		SO_3^{2-} sulfite (+4)	SO_4^{2-} sulfate (+6)
ClO^- hypochlorite (+1)	ClO_2^- chlorite (+3)	ClO_3^- chlorate (+5)	ClO_4^- perchlorate (+7)

We shall see many other examples of this tendency in later chapters.

1-6 CHEMICAL PROPERTIES SHOW TRENDS WITHIN EACH GROUP

In a broad sense, the chemical properties of the elements within a group are similar; yet there are certain anomalies and trends that occur within each group. Perhaps most important is the difference in the properties of the first member of a group from the others. Even in Group 1, which is a relatively homogeneous group, we shall find that lithium differs in a number of ways from the other members of the group. For example, most salts of sodium through cesium are soluble in water, but many lithium salts are only sparingly soluble. Lithium reacts directly with nitrogen at room temperature, but the other Group 1 metals react with nitrogen only at about 500°C. We shall see that the chemical properties of beryllium are different from those of magnesium and that boron (a semimetal) is quite different from aluminum. The most striking example of this difference, however, occurs in Group 4, where the chemistry of carbon is much different from that of the other Group 4 elements. The explanation of this anomaly is the very small ion involved (in the cases of lithium and beryllium), the lack of available *d* orbitals for bonding (in the case of carbon), the extreme electronegativity (in the cases of nitrogen, oxygen, and fluorine), or a combination of these factors.

The second-row elements have only 2*s* and 2*p* orbitals available for bonding, whereas the heavier elements also have *d* orbitals available. The heavier elements may "expand their shells" and accommodate more than eight electrons in their valence shells (see Section 10-10 of the text). Thus, for example, although NCl_3, PCl_3, and PCl_5 are known species, NCl_5 has never been observed. We shall see that this same effect can be used to account for the reactivity of $SiCl_4$, compared with the fairly unreactive CCl_4.

We also shall see that there is a surprising similarity in reactivity between lithium and beryllium, magnesium and aluminum, and boron and silicon. Some authors refer to these similarities as *diagonal relationships* because of the relative location of these pairs of elements in the periodic table (Figure 1-8).

There is also a curious reluctance for the fourth-row nonmetals (arsenic, selenium, and bromine) to achieve their maximum oxidation state. For example, in Group 5, arsenic pentoxide, As_4O_{10}, cannot be obtained from the reaction of arsenic with oxygen, although phosphorus pentoxide, P_4O_{10}, and antimony pentoxide, Sb_2O_5, can be prepared that way. Furthermore, As_4O_{10}, which can be prepared by other methods, readily loses oxygen when heated to give the trioxide, As_4O_6. In addition, $AsCl_5$ (as well as $AsBr_5$ and AsI_5) does not exist, whereas PCl_5 and $SbCl_5$ are stable. There is no satisfactory explanation for the reluctance of arsenic, selenium, and bromine to achieve their maximum oxidation states.

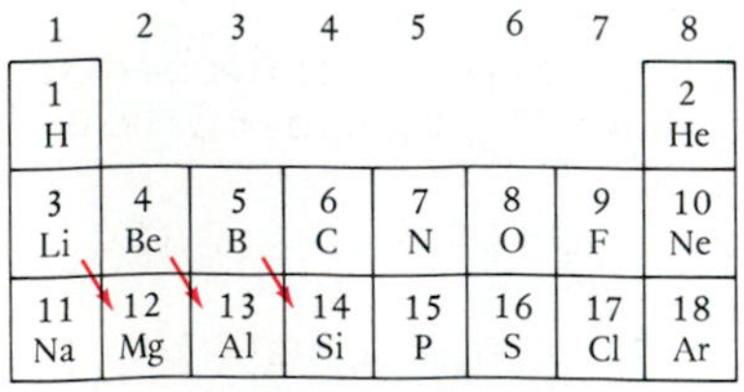

1	2	3	4	5	6	7	8
1 H							2 He
3 Li	4 Be	5 B	6 C	7 N	8 O	9 F	10 Ne
11 Na	12 Mg	13 Al	14 Si	15 P	16 S	17 Cl	18 Ar

Figure 1-8 Diagonal relationships among the *s*- and *p*-block elements in the periodic table. Note that the transition metals are not shown in the figure.

1-7 HIGHER OXIDATION STATES BECOME LESS STABLE ON DESCENDING A GROUP

The heavy post-transition metals (such as thallium, lead, and bismuth) have a strong tendency to occur in oxidation states that are two less than the highest possible oxidation state. For example, in Group 3, the +1 oxidation state of thallium

([Xe]$4f^{14}6s^2$) is more important than its +3 oxidation state ([Xe]$4f^{14}$). Indium displays a +1 oxidation state in a few compounds, whereas this state is unimportant for gallium. In Group 4, the only important oxidation state of germanium is +4; both the +2 and +4 oxidation states of tin are important; and the +2 oxidation state of lead is far more important than its +4 oxidation state. In Group 5, antimony occurs in the +3 and +5 oxidation states, but bismuth almost always occurs as bismuth(III). This effect is sometimes called the *inert s-pair effect,* an unfortunate name since it seems to imply a certain inertness of the 6*s* electrons in thallium, lead, and bismuth. However, there is no satisfactory explanation for this effect, and you should consider the name "inert *s*-pair effect" as a mnemonic rather than an explanation.

Now that we have discussed some general trends of chemical properties within the periodic table, we shall go on to study each group of elements in turn.

TERMS YOU SHOULD KNOW

malleable
ductile
semiconductor
s-block elements
p-block elements
d-block elements
f-block elements
main-group elements
transition metals
inner transition metals
lanthanides
actinides
period
electronegativity
bond enthalpy
diagonal relationship
inert *s*-pair effect

QUESTIONS

1-1. Without using your text, outline a periodic table and locate the *s*-block elements, *p*-block elements, *d*-block elements, and *f*-block elements.

1-2. How many elements are there in each *d*-transition series and in each *f*-transition series?

1-3. Why are the lanthanide series and the actinide series sometimes called the inner transition metal series?

1-4. List the key distinctions between metals, semimetals, and nonmetals.

1-5. Determine for each of the following elements whether they are metals, nonmetals, or semimetals.
(a) Li (b) Se (c) B (d) Cs (e) La (f) Mg (g) As (h) Si

1-6. Using only a periodic table, write the ground-state electron configurations of selenium, cesium, antimony, and indium.

1-7. Classify the following elements as *s*-block, *p*-block, *d*-block, or *f*-block elements.
(a) Ba (b) W (c) I (d) Te (e) Ga (f) Ar (g) Th (h) Sm

1-8. Based upon what you know from the periodic table, which of the following compounds would you expect not to exist?
(a) NCl_5 (b) NaO (c) Mn_2O_7 (d) CCl_6

1-9. A certain element, X, forms the following compounds: X_2O_3; XH_3; and XF_3. What are the possible group numbers for this element?

1-10. Sketch a periodic table, indicating the trend of electronegativities from highest to lowest. Do the same for atomic radii, metallic character, and ionization energy.

1-11. Determine for each of the following properties of elements whether the indicated property generally increases or decreases on moving from left to right across a row of the periodic table.

(a) metallic character
(b) electronegativity
(c) atomic size
(d) acidity of oxides
(e) ionization energy

1-12. Explain why the enthalpies of vaporization of the noble gases increase with increasing atomic number.

1-13. What is one of the most important differences between the bonding involving elements in the second row and the third row of the periodic table?

1-14. What is meant by "diagonal relationship" in the periodic table?

1-15. What is meant by the "inert *s*-pair effect"?

HYDROGEN

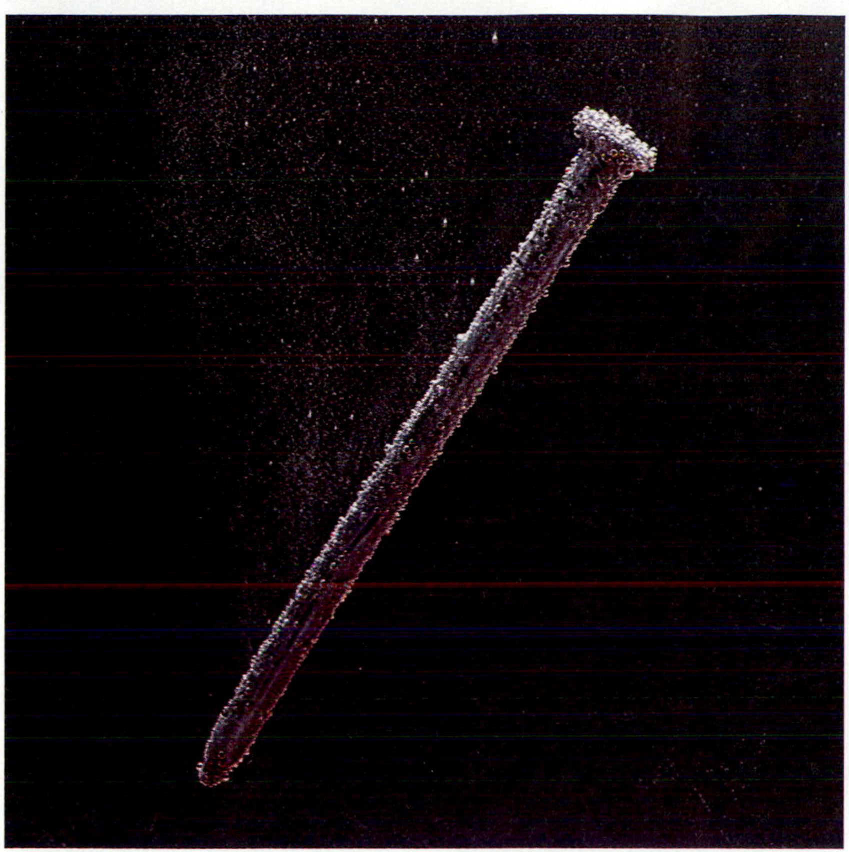

Iron metal reacts with dilute aqueous sulfuric acid to liberate hydrogen gas. Other reactive metals undergo similar reactions.

The ground-state electron configuration of atomic hydrogen is $1s^1$, so it is usually placed at the top of Group 1 in the periodic table because, like the Group 1 metals, it has a single electron in an outer *s* orbital. However, hydrogen is not a metal; rather, it exists as a diatomic gas under ordinary conditions. The oxidation states of hydrogen are +1, 0, and −1. The +1 oxidation state is exclusively covalent; the only ionic species is the −1 oxidation state of the *hydride ion,* H^-, which occurs in metallic hydrides formed from reactive metals and hydrogen. Because of its nonmetallic character and the existence of H^-, hydrogen is placed at the top of Group 7 in some versions of the periodic table.

Table 2-1 Isotopes of hydrogen

Isotope	*Atomic mass*/amu	*Natural abundance*/atom %
$^{1}_{1}H$	1.0078	99.985
$^{2}_{1}H$ (D)	2.0141	0.0148
$^{3}_{1}H$ (T)	3.0160	minute trace (about 1 in 10^{18} hydrogen atoms)

▪ On a mass basis hydrogen is the most exothermic fuel known.

Hydrogen gas has a lower molecular mass than that of any other gas and can be poured upward through the air from one container to another container. The low molecular mass of H_2 makes it the most effective gas for lighter-than-air balloons and aircraft. However, because hydrogen forms explosive mixtures with air, it is no longer used. Helium, being nonflammable, is now used for such applications. However, because of its low mass and its very exothermic reaction with oxygen, liquid hydrogen is used as a fuel in large booster rockets.

2-1 THERE ARE THREE HYDROGEN ISOTOPES

The element hydrogen, which has an atomic number of 1 and the lowest atomic mass (1.0079) of all the known elements, occurs as three different isotopes with mass numbers of 1, 2, and 3, respectively (Table 2-1). The mass 1 isotope is by far the most abundant of the three isotopes. The mass 2 isotope, which is called *deuterium,* is often denoted by the symbol D, and comprises only 1 out of about 6700 naturally occurring hydrogen atoms. Some of the physical properties of H_2 and D_2 are given in Table 2-2. The mass 3 isotope, which is called *tritium,* is often denoted by the symbol T, and comprises only about 1 out of 10^{18} hydrogen atoms. Tritium is radioactive and emits β particles (see Chapter 24 of the text).

$$^{3}_{1}H \rightarrow ^{3}_{2}He + ^{\ 0}_{-1}e \qquad t_{1/2} = 12.4 \text{ yr}$$

Naturally occurring tritium is produced in the upper atmosphere primarily by the reaction of cosmic ray neutrons with $^{14}_{7}N$:

$$^{14}_{7}N + \underset{\text{cosmic ray neutron}}{^{1}_{0}n} \rightarrow ^{3}_{1}H + ^{12}_{6}C$$

Tritium is produced for commercial use and in hydrogen bomb explosions by the reaction

$$^{6}_{3}Li + ^{1}_{0}n \rightarrow ^{3}_{1}H + ^{4}_{2}He$$

Atmospheric tritium levels increased over a hundredfold as the result of atmospheric nuclear tests during the 1950s.

Table 2-2 Physical properties of H_2 and D_2

Property	H_2	D_2
enthalpy of dissociation at 25°C/kJ · mol^{-1}	435.93	443.35
bond length/pm	74	74
melting point/K	14.0	18.7
boiling point/K	20.4	23.7
$\Delta\overline{H}^{\circ}_{vap}$/kJ · mol^{-1}	0.90	1.23
$\Delta\overline{H}^{\circ}_{fus}$/kJ · mol^{-1}	0.12	0.20
gas density at 0°C and 1 atm/g · L^{-1}	0.0899	0.180

2-2 HYDROGEN IS THE MOST ABUNDANT ELEMENT IN THE UNIVERSE

Under ordinary terrestrial conditions elemental hydrogen occurs as the colorless, odorless and tasteless diatomic gaseous molecule, H_2. Very little H_2 exists in the earth's crust or in the atmosphere, although some H_2 is found in volcanic gases and natural gas deposits. Hydrogen is rare in the atmosphere because it escapes the gravitational attraction of the earth. Most of the earth's hydrogen is combined with oxygen in water. Hydrogen also occurs frequently in combination with carbon, sulfur, and nitrogen and is a constituent of all plant and animal matter.

▪ Ten percent by mass of the human body is hydrogen.

▪ The name hydrogen, which was coined by Lavoisier, means water former.

On a mass basis, hydrogen is the ninth ranked element in the earth's crust, comprising about 0.9 percent by mass of the crust, which includes the oceans. However, on an atomic basis, hydrogen comprises about 15 percent of the atoms in the earth's crust and ranks third in atomic abundance behind oxygen and silicon. Over 30 percent of the mass of the sun is atomic hydrogen, and most of the atoms in interstellar space are hydrogen atoms. On both an atomic basis and a mass basis, hydrogen is by far the most abundant element in the universe.

2-3 HYDROGEN IS USUALLY PREPARED BY SINGLE-REPLACEMENT REACTIONS

Small quantities of hydrogen gas can be prepared in the laboratory by the reaction of zinc metal with aqueous hydrochloric acid (see frontispiece):

$$Zn(s) + 2HCl(aq) \rightarrow H_2(g) + ZnCl_2(aq)$$

Any metal whose standard reduction voltage E^0 for the half-reaction

$$M^{n+}(aq) + ne^- \rightarrow M(s)$$

is less than zero volts (see Table 21-1 of the text) is capable of liberating $H_2(g)$ from aqueous acids. Hydrogen can also be prepared by the reaction of calcium metal with cold water (Figure 2-1):

$$Ca(s) + 2H_2O(l) \rightarrow H_2(g) + Ca(OH)_2(aq)$$

Alexander Boden

Figure 2-1 Calcium metal reacts with cold water to produce hydrogen gas and calcium hydroxide.

The Group 1 metals also react with water to produce hydrogen, but these reactions are too violent to use for the routine preparation of hydrogen. For example, the reaction of potassium metal with water in the presence of air usually results in an explosion (Figure 2-2).

$$2K(s) + 2H_2O(l) \rightarrow H_2(g) + 2KOH(aq)$$

Alexander Boden

Figure 2-2 Potassium metal reacts explosively on contact with water.

The explosion occurs because the reaction of potassium with water is very fast and highly exothermic, and the resulting hot $H_2(g)$ reacts explosively with $O_2(g)$ in the air:

$$2H_2(g) + O_2(g) \rightarrow 2H_2O(g)$$

Hydrogen is prepared both commercially and for laboratory use by the electrolysis of aqueous solutions of sulfuric acid using either platinum or nickel electrodes (Figure 2-3).

$$2H_2O(l) \xrightarrow{\text{electrolysis}} 2H_2(g) + O_2(g)$$

The role of the sulfuric acid is that of an electrolyte, carrying the ionic current through the solution and thereby decreasing the resistance of the solution. Numerous other electrolytes with nonoxidizable anions could be used. The half-reactions occurring at the cathode and the anode in the electrolysis of an $H_2SO_4(aq)$ solution are

$$4H^+(aq) + 4e^- \rightarrow 2H_2(g) \qquad \text{(cathode, reduction)}$$

$$2H_2O(l) \rightarrow 4H^+(aq) + 4e^- + O_2(g) \qquad \text{(anode, oxidation)}$$

The sum of the two half-reactions gives the overall net reaction.

■ Over 200 million pounds of hydrogen are produced annually in the United States.

The major industrial methods for the preparation of hydrogen involve endothermic reactions of high-temperature steam with hydrocarbons from natural gas or oil-refinery sources using nickel catalysts. For example,

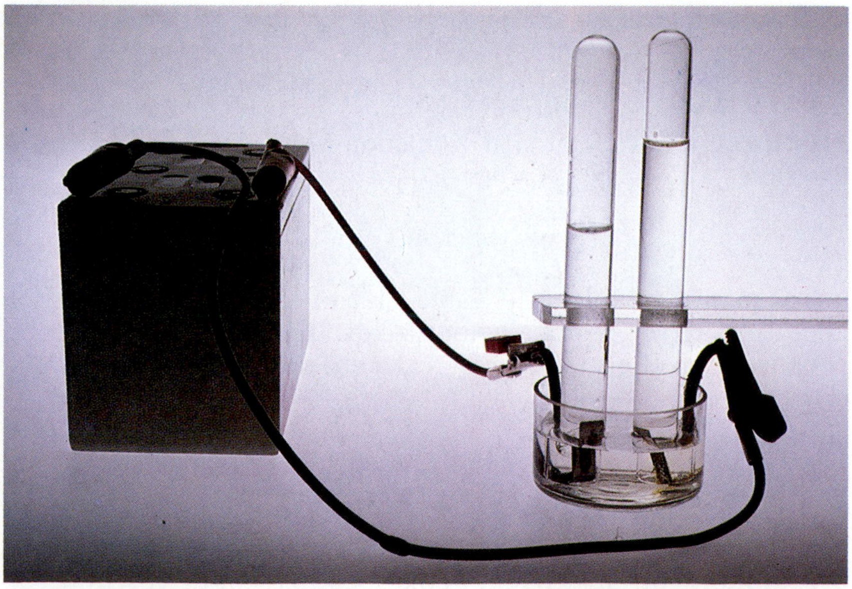

Figure 2-3 Preparation of hydrogen by electrolysis on an aqueous sulfuric acid solution. Hydrogen gas is liberated at the cathode, and oxygen gas is liberated at the anode. Note, as is required by the reaction stoichiometry, that the volume of $H_2(g)$ liberated is twice as great as that of $O_2(g)$.

$$CH_4(g) + H_2O(g) \xrightarrow[1000°C]{Ni} CO(g) + 3H_2(g) \qquad \Delta H^\circ_{rxn} = +206 \text{ kJ at } 25°C$$

Another important industrial process for hydrogen production is the *water-gas reaction:*

$$C(\text{coke}) + H_2O(g) \xrightarrow[1000°C]{\text{Fe or Ru}} CO(g) + H_2(g) \qquad \Delta H^\circ_{rxn} = +130 \text{ kJ at } 25°C$$

The coke is obtained by heating coal in the absence of oxygen to vaporize volatile constituents of the coal. In both of these processes, additional hydrogen is generated by reacting the carbon monoxide with additional steam at about 400°C over an iron oxide catalyst.

$$CO(g) + H_2O(g) \xrightarrow[400°C]{Fe_2O_3} CO_2(g) + H_2(g) \qquad \Delta H^\circ_{rxn} = -41.2 \text{ kJ at } 25°C$$

2-4 THE H_2 MOLECULE IS FAIRLY UNREACTIVE BECAUSE OF THE STRONG BOND

The molar enthalpy of dissociation of H_2 is 436 $kJ \cdot mol^{-1}$ at 25°C:

$$H_2(g) \rightarrow 2H(g) \qquad \Delta H^\circ_{rxn} = 436 \text{ kJ} \cdot \text{mol}^{-1}$$

and thus the single sigma bond ($1\sigma^2$) in H_2 is a relatively strong bond. The strong H—H bond makes H_2 a fairly unreactive molecule, because this bond must be broken in order for H_2 to react. The lack of reactivity of H_2 can be illustrated by the following data (25°C):

$$H_2(g) + \tfrac{1}{2}O_2(g) \rightarrow H_2O(g) \qquad \Delta H^\circ_{rxn} = -242 \text{ kJ} \quad \Delta G^\circ_{rxn} = -229 \text{ kJ}$$

$$3H_2(g) + 2N_2(g) \rightarrow 2NH_3(g) \qquad \Delta H^\circ_{rxn} = -92 \text{ kJ} \quad \Delta G^\circ_{rxn} = -33 \text{ kJ}$$

In both cases the reaction products are thermodynamically favored under standard conditions ($\Delta G^\circ_{rxn} < 0$), but if we prepare the reaction mixtures at 25°C, the reactions do not occur at a measurable rate. Although the $H_2 + O_2$ reaction can be initiated by a spark, most reactions involving H_2 require the presence of catalysts. For example, many reactions involving H_2 are catalyzed by platinum and palladium. These catalysts work by facilitating the dissociation of H_2 into H atoms on the surface of the metal:

$$H_2(g) \xrightarrow{Pt} 2H(\text{surface})$$

where the surface H atoms are attached to Pt atoms on the metal surface.

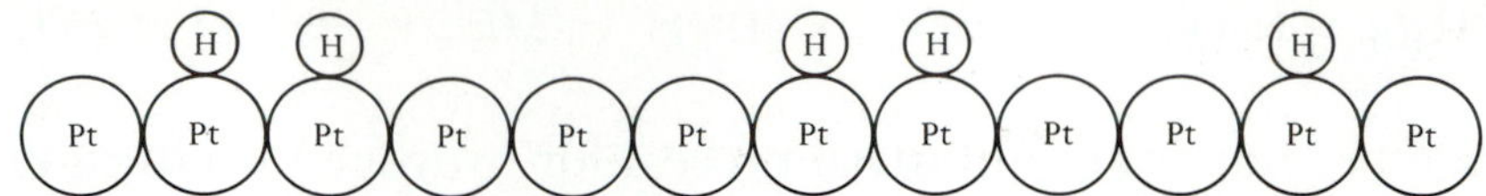

Science Museum, London

When the catalytic effect of platinum metal on the combustion of hydrogen was discovered in the early 1800s, the process was used to produce cigar lighters, which became very fashionable.

A dramatic illustration of the catalytic activity of platinum metal on hydrogen can be seen by passing hydrogen plus air over finely divided platinum. At room temperature, the platinum will glow sufficiently to ignite the hydrogen. When this process was first discovered in the early 1800s, it was used to produce lighters.

2-5 THE MAJOR INDUSTRIAL USE OF HYDROGEN IS IN THE SYNTHESIS OF AMMONIA

Ammonia is produced commercially by the *Haber process,* which involves the reaction

$$3H_2(g) + N_2(g) \xrightarrow[300\text{ atm, }500^\circ\text{C}]{\text{Fe + Mo}} 2NH_3(g)$$

The activation energy for this reaction is 336 $kJ \cdot mol^{-1}$ in the absence of the catalyst and about 150 $kJ \cdot mol^{-1}$ in the presence of the catalyst. About 32 billion pounds of ammonia are produced annually in the United States via the above reaction (see Section 15-11 of the text for a more detailed discussion of the Haber process).

Large quantities of hydrogen are used in the platinum-and-nickel-catalyzed hydrogenation of unsaturated liquid vegetable oils to produce saturated solid fats, such as margarine. These addition reactions are of the type

$$\begin{matrix} R & & R' \\ & C{=}C & \\ H & & H \end{matrix}(l) + H_2(g) \xrightarrow{\text{Pt or Ni}} R{-}\overset{\displaystyle H}{\underset{\displaystyle H}{C}}{-}\overset{\displaystyle H}{\underset{\displaystyle H}{C}}{-}R'(s)$$

where R and R′ are hydrocarbon segments. Hydrogen is also used on a large scale in the manufacture of various organic chemicals, especially methanol by the reaction

$$2H_2(g) + CO(g) \xrightarrow{\text{cobalt}} CH_3OH(l) \qquad \Delta G^\circ_{rxn} = -29\text{ kJ at }25^\circ\text{C}$$

Hydrogen is also used in metallurgy to reduce metal oxides such as those of molybdenum and tungsten to the metal. For example,

$$MoO_3(s) + 3H_2(g) \xrightarrow[600^\circ\text{C}]{} Mo(s) + 3H_2O(g)$$

The reaction

$$2H_2(g) + O_2(g) \rightarrow 2H_2O(g)$$

is used in the oxy-hydrogen torch. The oxy-hydrogen torch has a flame temperature of about 2500°C and is used in cutting and welding metals. The explosive reaction of the gases is prevented by mixing them just before they reach the burner orifice.

Temperatures of up to 4000°C are produced in the atomic hydrogen torch, where hydrogen atoms are generated in an electric arc and are allowed to recombine on a metal surface to produce H_2. The energy liberated by bond formation produces very high temperatures, which can be used to weld high-melting metals such as tungsten and tantalum.

An oxy-hydrogen torch. The flame temperature is about 2500°C.

2-6 METAL HYDRIDES ARE FORMED BY THE DIRECT REACTION OF ACTIVE METALS WITH H_2

The very reactive Group 1 and Group 2 metals react directly with H_2 to produce metal hydrides, which are white ionic crystals that contain the hydride ion, H^-.

$$2Na(s) + H_2(g) \rightarrow \underset{\text{sodium hydride}}{2NaH(s)}$$

$$Ca(s) + H_2(g) \rightarrow \underset{\text{calcium hydride}}{CaH_2(s)}$$

The reactions of the Group 2 metals with hydrogen are more vigorous than those of the Group 1 metals; for example, beryllium and magnesium burst into flame on reaction with hydrogen. The Group 1 and Group 2 metal hydrides are powerful reducing agents that liberate hydrogen from water; for calcium hydride,

$$CaH_2(s) + 2H_2O(l) \rightarrow H_2(g) + Ca(OH)_2(aq)$$

Calcium hydride (CaH_2), a white crystalline ionic solid, reacts readily with water to produce hydrogen gas.

Some transition-metal hydrides have well-defined stoichiometry—for example, $TlH_2(s)$, $CuH(s)$, $CeH_2(s)$, and $CeH_3(s)$—but many others are *nonstoichiometric compounds,* in which the hydrogen atoms occupy cavities in the crystal lattice. Nonstoichiometric compounds are compounds in which the elements are not combined in definite small whole-number ratios. For example, hydrogen reacts with palladium to form a substance of the composition PdH_x, where the observed value of x ranges from zero to somewhat less than 1. The fact that x can have a continuous range of values is in sharp contrast to that of NaH, where the Na/H ratio is one.

Hydrogen forms more compounds than any other element. Compounds of hydrogen with the nonmetals are covalently bonded low-boiling, molecular compounds. These compounds are discussed under the individual groups of the elements.

▪ Nonstoichiometric compounds are called berthollide compounds.

TERMS YOU SHOULD KNOW

hydride ion
deuterium
tritium
water-gas reaction
Haber process
nonstoichiometric compound
berthollide compound

QUESTIONS

2-1. Using the data in Table 2-1, calculate the atomic mass of naturally occurring hydrogen.

2-2. Suggest how tritium can be used to study the movement of ground water.

2-3. Calculate the number of disintegrations per second in a sample consisting of one micromole of tritium.

2-4. A sample of water containing a trace amount of T_2O is found to have 1.12×10^4 disintegrations per second. What will be its activity after 50 years?

2-5. Complete and balance the following equations.

(a) $Fe_2O_3(s) + H_2(g) \xrightarrow{\text{high T}}$

(b) $LiH(s) + H_2O(l) \rightarrow$

(c) $Mg(s) + H_2(g) \rightarrow$

(d) $K(s) + H_2(g) \rightarrow$

(e) $H_2(g) + H_2C{=}CH_2(g) \xrightarrow{\text{Pt}}$

2-6. Complete and balance the following equations.

(a) $Zn(s) + HBr(aq) \rightarrow$

(b) $C(s) + H_2O(g) \xrightarrow[1000°C]{\text{Fe}}$

(c) $D_2(g) + N_2(g) \xrightarrow{\text{Fe/Mo}}$

(d) $Li(s) + D_2O(l) \rightarrow$

(e) $W(s) + H_2O(g) \xrightarrow{\text{high T}}$

2-7. In the Three Mile Island nuclear power plant accident, a large quantity of hydrogen gas was produced in the high temperature reaction between steam and the zirconium metal in the fuel rod assemblies. Given that ZrO_2 is the formula for the zirconium oxide formed, write a balanced equation for the reaction.

2-8. Complete and balance the following equation:

$$C_3H_8(g) + H_2O(g) \xrightarrow[1000°C]{\text{Ni}}$$

2-9. Lithium metal is often cleaned by treatment with ethanol, C_2H_5OH. By analogy with the reaction of Li(s) with water, complete and balance the following equation:

$$Li(s) + CH_3CH_2OH(l) \rightarrow$$

2-10. Hydrogen is the most exothermic fuel on a mass basis. Suggest some reasons why hydrogen is not a widely used fuel.

2-11. How many grams of zinc are required to generate 500 mL of hydrogen at 20°C and 740 torr by the reaction between zinc and hydrochloric acid?

2-12. Calculate the relative masses of hydrogen produced by the reaction of hydrochloric acid with 100 g of iron and 100 g of zinc.

2-13. What volume of hydrogen at 250°C and 10.0 atm is required to reduce 2.50 metric tons of tungsten (VI) oxide to the metal?

2-14. Calculate the enthalpy of combustion of one gram of hydrogen.

2-15. Given that $\Delta G^\circ_{rxn} = -191$ kJ at 25°C for the reaction

$$H_2(g) + Cl_2(g) \rightleftharpoons 2HCl(g)$$

calculate the maximum voltage that can be obtained from a fuel cell utilizing this reaction with each gas at 1.00 atm pressure.

2-16. Given that $\Delta G^\circ_{rxn} = -237$ kJ at 25°C for the reaction

$$H_2(g) + \tfrac{1}{2}O_2(g) \rightleftharpoons H_2O(l)$$

calculate the minimum voltage required to decompose water electrolytically.

THE ALKALI METALS

The reaction of sodium with water. The piece of molten sodium is propelled along the water surface by the evolved hydrogen. The production of NaOH(*aq*) is shown by the pink color of the acid-base indicator phenolphthalein, which is colorless in acidic or neutral solution and pink in basic solution. The heat of the reaction is sufficient to melt the sodium.

The Group 1 elements, lithium, sodium, potassium, rubidium, cesium, and francium, are reactive metals with electron configurations of the type [noble gas]ns^1. These elements attain a noble-gas electron configuration by the loss of one electron.

$$\text{M}\{[\text{noble gas}]ns^1\} \rightarrow \text{M}^+[\text{noble gas}] + e^-$$

Because of their relatively low first-ionization energies and high second ionization energies, the chemistry of these elements involves primarily the metals and the +1 ions. As a family, the Group 1 metals show clearly the effect of increasing atomic number on physical and chemical properties. Their atomic and ionic radii increase uniformly and their ionization energies decrease uniformly with increasing atomic number. With few exceptions, the reactivity of the Group 1 elements increases from lithium to cesium. As in all the *s*-block and *p*-block groups, the first member of a family differs in a number of respects from the other members. For example, although most salts of the Group 1 metals are soluble in water, a number of lithium salts (LiOH, LiF, Li_2CO_3) are only sparingly soluble. The anomalous properties of lithium can be attributed to its rather small ionic radius (76 pm). In fact, the radius of Li^+ is similar to that of Mg^{2+} (72 pm), and lithium has some chemical properties similar to those of magnesium. Many magnesium salts are less water soluble than the heavier members of Group 2. The similarity between lithium and magnesium is an example of a diagonal relationship between elements in the periodic table.

3-1 THE GROUP 1 ELEMENTS DO NOT OCCUR AS THE FREE METAL IN NATURE

William Garnett

Figure 3-1 Owens Lake, a brine lake in California near Death Valley.

Because the Group 1 metals have relatively low ionization energies, they are all very reactive and do not occur as the free metal in nature. Lithium is a fairly rare element, occurring in the earth's crust to the extent of about 20 ppm (parts per million) by mass. Because of its chemical similarity to magnesium, lithium is found associated with several magnesium minerals. The most important ore of lithium is *spodumene,* $LiAlSi_2O_6$, large deposits of which occur in South Dakota, Manitoba, the U.S.S.R., and Brazil. Lithium salts also occur in certain brine lakes in California and Nevada (Figure 3-1).

Sodium and potassium are the sixth and seventh most abundant elements, respectively, in the earth's crust. Vast deposits of the salts of both metals have resulted from the evaporation of ancient seas. Although a number of salts serve as commercial sources of these two metals, NaCl (*halite*) and KCl (*sylvite*) are the most important. Both rubidium and cesium occur in small quantities with the other alkali metals. There are no stable isotopes of francium; the longest-lived isotope is ^{223}Fr, with a half life of only 22 minutes. Table 3-1 summarizes the principal commercial sources and uses of the alkali metals.

3-2 THE PROPERTIES OF THE GROUP 1 METALS DEPEND UPON THE SIZE OF THE ATOMS

The Group 1 metals are all fairly soft and can be cut with a sharp knife (Figure 3-2). When freshly cut they are bright and shiny, but they soon take on a dull finish because of the reactions with

Table 3-1 The sources and uses of the alkali metals

Metal	*Sources*	*Uses*
lithium	spodumene, $LiAlSi_2O_6(s)$; certain mineral springs and salt lakes	alloys, organic reactions, degasification of copper
sodium	salt waters, $NaCl(s)$, $NaNO_3(s)$	synthesis of tetraethyllead, production of titanium metal, small nuclear reactor coolant
potassium	ancient ocean and salt lake beds; occurs in numerous mineral deposits at low levels, $KNO_3(s)$, $KCl(s)$	heat exchange alloys
rubidium	mineral springs (Searles Lake, Calif.; Manitoba; Michigan brines), certain rare minerals found in Elba	photocells; getter (O_2 remover) in vacuum tubes
cesium	water from certain mineral springs (Bernic Lake, Manitoba),	ion propulsion systems; atomic clocks

air (Figure 3-2). They must be stored under an inert substance such as kerosene, because they react spontaneously with oxygen and water vapor in air. The Group 1 metals are also called the *alkali metals,* because their hydroxides, MOH, are all soluble, strong bases in water (alkaline means basic). The atomic and

Figure 3-2 The Group 1 metals are soft. Here we see sodium being cut with a knife. Note the shiny surface of the freshly cut metal and the tarnished surface that results on exposure to air.

Table 3-2 The atomic properties of the Group 1 elements

Property	*Lithium*	*Sodium*	*Potassium*	*Rubidium*	*Cesium*	*Francium*
chemical symbol	Li	Na	K	Rb	Cs	Fr
atomic number	3	11	19	37	55	87
atomic mass	6.941	22.98977	39.0983	85.4678	132.9054	(223)
number of naturally occurring isotopes	2	1	3	2	1	0
ground-state electron configuration	$[He]2s^1$	$[Ne]3s^1$	$[Ar]4s^1$	$[Kr]5s^1$	$[Xe]6s^1$	$[Rn]7s^1$
metal radius/pm	145	180	220	235	266	~290
ionic radius/pm	60	95	133	148	169	~185
first ionization energy of M(*g*)/ $kJ \cdot mol^{-1}$	520	496	419	403	376	~370
Pauling electronegativity	1.0	0.9	0.8	0.8	0.7	0.7

physical properties of the alkali metals are given in Tables 3-2 and 3-3, respectively.

The periodic trends of the alkali metals are easily seen in the data given in Tables 3-2 and 3-3. The first ionization energy and the electronegativity decrease as we go down the group, whereas the metal radius and ionic radius increase (Table 3-2). These trends are a direct consequence of the increase in size of the atoms resulting from the increase in the number of electrons with increasing atomic number.

▪ Lithium is the least dense of all the elements that are solid or liquid at 20°C.

In Table 3-3 we note that as we descend the group, there is a decrease in the melting and boiling points and in the enthalpies of fusion and vaporization. All these decreases are a result of the increasing size of the alkali metals as we move down the group. The increase in density as we descend the group is a consequence of the increase in atomic mass.

Table 3-3 The physical properties of the Group 1 elements

Property	*Lithium*	*Sodium*	*Potassium*	*Rubidium*	*Cesium*
melting point/°C	181	98	64	39	29
boiling point/°C	1347	892	774	696	670
density at 20°C/$g \cdot cm^{-3}$	0.53	0.97	0.86	1.53	1.88
$\Delta\overline{H}_{fus}/kJ \cdot mol^{-1}$	2.93	2.64	2.39	2.20	2.09
$\Delta\overline{H}_{vap}/kJ \cdot mol^{-1}$	148	99	79	76	67
E^0/V at 25°C for $M^+(aq) + e^- \rightarrow M(s)$	−3.05	−2.71	−2.93	−2.93	−2.92

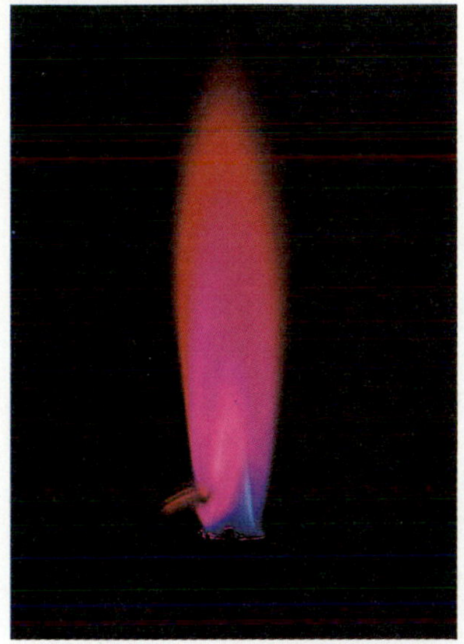

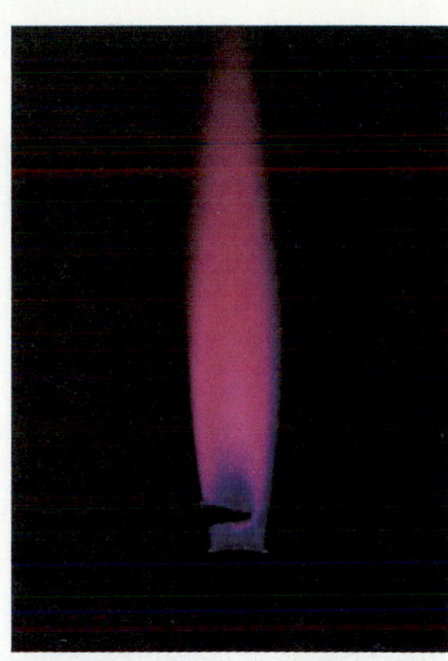
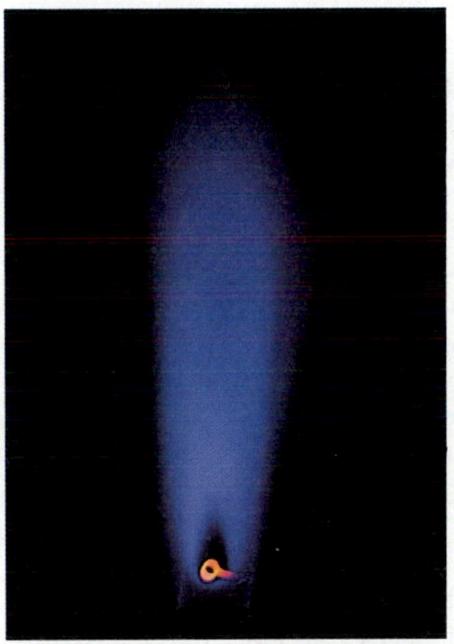
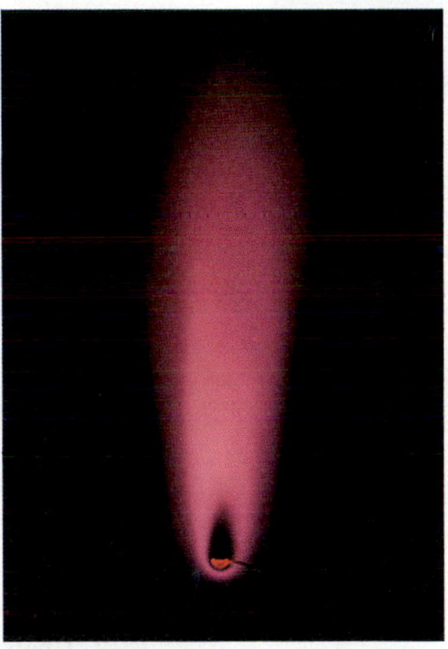

Flames of the Group 1 metals. In the top row from the left: lithium (crimson), sodium (yellow), and potassium (violet); in the second row, rubidium (blue) and cesium (pale violet). The colors, which arise from electronic transitions in the electronically excited metal atoms, are used in qualitative analysis to detect the presence of alkali metal ions in a sample. Ions are reduced to the gaseous metal atoms in the lower central region of the flame.

3-3 THE ALKALI METALS CAN BE OBTAINED BY ELECTROLYSIS OF THE MOLTEN CHLORIDES

Sodium metal is obtained by electrolysis of molten mixtures of sodium chloride and calcium chloride:

$$2NaCl\ [\text{in } CaCl_2(l)] \xrightarrow[600°C]{\text{electrolysis}} 2Na(l) + Cl_2(g)$$

Chlorine gas is a useful by-product of the electrolysis. The $CaCl_2$ is added to the NaCl to lower the temperature necessary for the operation of the electrolysis cell. Pure NaCl melts at 800°C.

Potassium and the other alkali metals also can be obtained by electrolysis. An alternate preparation of, for example, potassium involves the replacement reaction of molten potassium chloride with gaseous sodium in the absence of air.

$$KCl(l) + Na(g) \xrightarrow{780^\circ C} NaCl(s) + K(g)$$

The success of the process is based on the fact that potassium is much more volatile than sodium. The boiling point of potassium is 118°C lower than that of sodium (Table 3-3). Rubidium and cesium can be produced in an analogous manner.

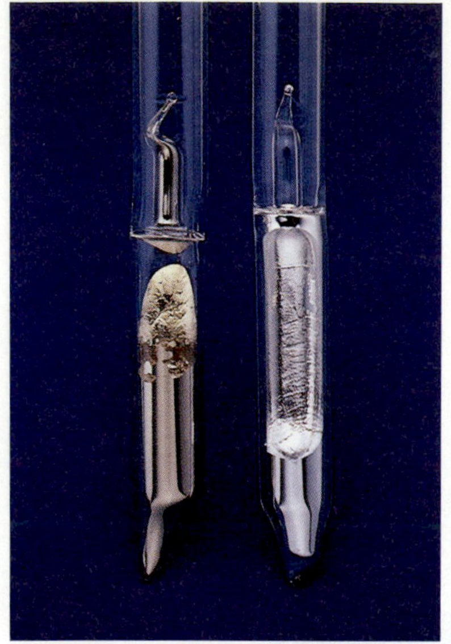

Cesium (which is a gold color) and rubidium (silver). They are stored in vacuum tubes to prevent them from reacting with the air.

3-4 GROUP 1 COMPOUNDS ARE GENERALLY IONIC, WATER-SOLUBLE SALTS

The alkali metals react directly with all the nonmetals except the noble gases (Table 3-4). The increasing reactivity of the alkali metals with increasing atomic number is demonstrated in a spectacular manner by their reaction with water. When metallic lithium reacts with water, hydrogen gas is slowly evolved, whereas sodium reacts vigorously with water (see frontispiece). The reaction of potassium with water produces a fire (Figure 2-2) because the heat generated by the reaction is sufficient to ignite the hydrogen gas evolved. Rubidium and cesium react with water with explosive violence.

Molten lithium is an exceedingly reactive substance. The only known substances that do not react with molten lithium are tungsten, molybdenum, and low-carbon stainless steels. If a piece of lithium metal is melted in a glass tube, then the molten lithium rapidly eats a hole through the glass. The reaction is accompanied by a brilliant green-yellow flame and considerable evolution of heat.

The alkali metals react directly with oxygen. Molten lithium ignites in oxygen to form $Li_2O(s)$; the reaction is accompanied by

Table 3-4 Some of the more common reactions of the alkali metals

reaction with oxygen $4Li(s) + O_2(g) \rightarrow 2Li_2O(s)$ $2Na(s) + O_2(g) \rightarrow Na_2O_2(s)$ $K(s) + O_2(g) \rightarrow KO_2(s)$ $Cs(s) + O_2(g) \rightarrow CsO_2(s)$ $Rb(s) + O_2(g) \rightarrow RbO_2(s)$	reaction with water $2M(s) + 2H_2O(l) \rightarrow 2MOH(s) + H_2(g)$ reaction with sulfur $2M(s) + S(s) \rightarrow M_2S(s)$ reaction with hydrogen $2M(s) + H_2(g) \xrightarrow{500\text{–}800^\circ C} 2MH(s)$
reaction[a] with halogens (denoted by X_2) $2M(s) + X_2 \rightarrow 2MX(s)$	reaction with nitrogen[b] $6M(s) + N_2(g) \xrightarrow{600^\circ C} 2M_3N(s)$

[a] M(s) denotes any one of the alkali metals.
[b] Li(s) reacts with N_2 at room temperature.

a bright red flame. The reactions of the other alkali metals do not yield the oxides M_2O. With sodium the *peroxide* Na_2O_2 is formed, and with potassium, rubidium, and cesium the *superoxides* KO_2, RbO_2, and CsO_2 are formed.

Both sodium peroxide and potassium superoxide are used in self-contained breathing apparatus. In the case of Na_2O_2, the relevant reaction is

$$2Na_2O_2(s) + \underset{\text{exhaled air}}{2CO_2(g)} \rightarrow 2Na_2CO_3(s) + O_2(g)$$

and in the case of $KO_2(s)$, there are two key reactions:

$$4KO_2(s) + \underset{\text{exhaled air}}{2H_2O(g)} \rightarrow 3O_2(g) + 4KOH(s)$$

$$KOH(s) + CO_2(g) \rightarrow KHCO_3(s)$$

The alkali metals react directly with hydrogen at high temperatures to form hydrides. For example,

$$2Na(l) + H_2(g) \xrightarrow{500°C} 2NaH(s)$$

The alkali metal hydrides are ionic compounds that contain the hydride ion, H^-. The hydrides react with water to liberate hydrogen,

$$NaH(s) + H_2O(l) \rightarrow NaOH(aq) + H_2(g)$$

and are used to remove traces of water from organic solvents. In such cases, the metal hydroxide precipitates from the solution.

Lithium is the only element that reacts directly with nitrogen at room temperature:

$$6Li(s) + N_2(g) \rightarrow \underset{\text{lithium nitride}}{2Li_3N(s)}$$

Lithium nitride, Li_3N.

The other alkali metals react with $N_2(g)$ at higher temperatures.

Compounds of the alkali metals are for the most part white, high-melting ionic solids. With very few exceptions, alkali metal salts are soluble in water and the resulting solutions are electrolytic as a result of the dissociation of the salt into its constituent ions. As noted earlier, some lithium salts are insoluble in water.

The alkali metals have the unusual property of dissolving in liquid ammonia to yield a blue electrolytic solution. The properties of such a solution are interpreted in terms of *solvated electrons* and alkali metal ions:

$$M(s) \xrightarrow[NH_3(l)]{} M^+(\text{amm}) + e^-(\text{amm})$$

When the blue solutions are concentrated by evaporation, they become bronze in color and behave like liquid metals.

▪ Anions of the alkali metals (alkalide ions) can be prepared by dissolving the metal in ethylenediamine, $H_2NCH_2CH_2NH_2$, in the presence of certain organic chelating agents; on cooling, a salt of the type $Na^+(\text{chelate})Na^-$ forms, which involves the alkalide ion Na^-.

3-5 MANY ALKALI METAL COMPOUNDS ARE IMPORTANT COMMERCIALLY

Sodium hydroxide is the seventh ranked industrial chemical. Over 20 billion pounds of it is produced annually in the United States. Sodium hydroxide sometimes is called *caustic soda* and is prepared by the electrolysis of concentrated aqueous sodium chloride solutions:

$$2NaCl(aq) + 2H_2O(l) \xrightarrow{\text{electrolysis}} 2NaOH(aq) + H_2(g) + Cl_2(g)$$

or by the reaction between calcium hydroxide (called slaked lime) and sodium carbonate:

$$Na_2CO_3(aq) + Ca(OH)_2(aq) \rightarrow 2NaOH(aq) + CaCO_3(s)$$

The formation of the insoluble $CaCO_3$ is a driving force for this second reaction. The alkali metal hydroxides are white, translucent, corrosive solids that are extremely soluble in water; at 20°C the solubility of NaOH is 15 M and that of KOH is 13 M.

▪ About 90 percent of the soda ash produced in the U.S. is obtained from natural deposits of the mineral *trona,* which is $Na_2CO_3 \cdot NaHCO_3 \cdot 2H_2O(s)$.

Sodium carbonate, which is called *soda ash,* is the tenth ranked industrial chemical. The annual United States production of sodium carbonate exceeds 16 billion pounds. It is prepared from sodium chloride by the *Solvay process,* which was devised by the Belgian brothers Ernest and Edward Solvay in 1861. In this process, carbon dioxide is bubbled through a cooled solution of sodium chloride and ammonia. The reactions are

$$NH_3(aq) + CO_2(aq) + H_2O(l) \rightarrow NH_4^+(aq) + HCO_3^-(aq)$$

$$NaCl(aq) + NH_4^+(aq) + HCO_3^-(aq) \xrightarrow{15°C} NaHCO_3(s) + NH_4Cl(aq)$$

At 15°C the sodium hydrogen carbonate precipitates from the solution. Part of the sodium hydrogen carbonate is converted to sodium carbonate by heating:

$$2NaHCO_3(s) \xrightarrow{80°C} Na_2CO_3(s) + H_2O(l) + CO_2(g)$$

The carbon dioxide produced in this reaction is used again in the first reaction.

The commercial success of the Solvay process requires the recovery of the ammonia, which is relatively expensive. The ammonia is recovered from the NH_4Cl by the reaction

$$2NH_4Cl(aq) + Ca(OH)_2(s) \rightarrow 2NH_3(g) + CaCl_2(aq) + 2H_2O(l)$$

The calcium hydroxide and the carbon dioxide used in the process are obtained by heating limestone (primarily $CaCO_3$).

The raw materials of the Solvay process are ammonia, sodium chloride, limestone, and water. The ammonia is recovered, and the other three substances are inexpensive. The principal use of sodium carbonate is in the manufacture of glass.

Some other important alkali metal compounds and their uses are given in Table 3-5.

Table 3-5 Some commercially important alkali metal compounds and their uses

Compound	*Uses*
lithium aluminum hydride, $LiAlH_4(s)$	production of many pharmaceuticals, perfumes, and organic chemicals
lithium borohydride, $LiBH_4(s)$	strong reducing agent, used in organic synthesis
lithium carbonate, $Li_2CO_3(s)$	to treat schizophrenia
lithium fluoride, $LiF(s)$	flux for aluminum soldering and welding; large crystals used as prisms in infrared spectrometers
sodium hydrogen carbonate (sodium bicarbonate), $NaHCO_3(s)$	manufacture of effervescent salts and beverages, baking powder, gold plating
sodium carbonate, $Na_2CO_3(s)$	manufacture of glass, pulp and paper, soaps and detergents, textiles
sodium hydroxide, $NaOH(s)$	production of rayon, cellulose, paper, soaps, detergents, textiles, oven cleaner
sodium sulfate decahydrate (Glauber's salt), $Na_2SO_4 \cdot 10H_2O(s)$	solar heating storage, air conditioning
sodium bisulfite, $NaHSO_3(s)$	disinfectant and bleaching agent
sodium and potassium chlorite, $NaClO_3(s)$ and $KClO_3(s)$	explosives, fireworks, and matches; weed killer
sodium cyanide, $NaCN(s)$	extraction of gold and silver from ores; electroplating solutions; fumigant for fruit trees
sodium peroxide, $Na_2O_2(aq)$	bleaching agent, air purifier
potassium carbonate (potash), $K_2CO_3(s)$	manufacture of special glass for optical instruments and electronic devices, soft soaps
potassium nitrate, $KNO_3(s)$	pyrotechnics, explosives, matches; tobacco treatment
potassium permanganate(s), $KMnO_4$	decolorizer, bleaching agent, manufacture of saccharine
dipotassium hydrogen phosphate(s), K_2HPO_4	buffer agent
cesium bromide, $CsBr(s)$, and cesium iodide, $CsI(s)$	spectrometer prisms, fluorescent screens
cesium chloride, $CsCl(s)$	brewing, mineral waters

TERMS YOU SHOULD KNOW

spodumene
halite
sylvite
alkali metals
peroxide
superoxide
caustic soda
soda ash
Solvay process

QUESTIONS

3-1. Why must the alkali metals be stored under kerosene?

3-2. Explain why the reactivities of the alkali metals increase with atomic number.

3-3. How is sodium metal produced commercially?

3-4. What are the raw materials in the Solvay process? Write chemical equations for the reactions in this process.

3-5. Complete and balance the following equations.

(a) $Na(s) + H_2O(l) \rightarrow$

(b) $K(s) + Br_2(l) \rightarrow$

(c) $Li(s) + N_2(g) \rightarrow$

(d) $Na(s) + H_2(g) \xrightarrow{600°C}$

(e) $NaH(s) + H_2O(l) \rightarrow$

3-6. Complete and balance the following equations.

(a) $Li(s) + O_2(g) \rightarrow$

(b) $Na(s) + O_2(g) \rightarrow$

(c) $K(s) + O_2(g) \rightarrow$

(d) $Cs(s) + O_2(g) \rightarrow$

3-7. Complete and balance the following equations

(a) $KO_2(s) + H_2O(g) \rightarrow$

(b) $Na_2O_2(s) + CO_2(g)$

(c) $NaOH(s) + CO_2(g) \rightarrow$

(d) $NaNH_2(s) + H_2O(l) \rightarrow$

3-8. The sodium-sulfur battery has been extensively studied as a potential power source for electric powered vehicles. This high-temperature battery uses the elements sodium and sulfur in molten form. Write the anode and cathode half-reactions and the net cell reaction on discharge.

3-9. Explain why sodium metal cannot be prepared by the electrolysis of an $NaCl(aq)$ solution.

3-10. Solutions of sodium metal in liquid ammonia decompose in the presence of a rusty nail, liberating hydrogen gas and forming a white precipitate. Postulate a balanced chemical equation to explain these observations. Formulate your answer by analogy with the reaction between sodium metal and water.

3-11. Sodium peroxide is prepared by first oxidizing sodium to Na_2O in a limited supply of $O_2(g)$ and then reacting this further to give Na_2O_2. Why can't the peroxides of potassium, rubidium, and cesium be prepared in this manner?

3-12. Use molecular orbital theory (Section 12-13 in the text) to argue that the superoxide ion is paramagnetic, and that the peroxide ion is diamagnetic (not paramagnetic).

3-13. Sodium hydride is used to extract titanium from $TiCl_4$ according to

$$TiCl_4(g) + 4NaH(s) \xrightarrow{400°C} Ti(s) + 4NaCl(s) + 2H_2(g)$$

How many grams of NaH are required to react with one kilogram of $TiCl_4$?

3-14. Sodium hydride reacts with SO_2 to produce sodium dithionite, $Na_2S_2O_4$, according to

$$2SO_2(l) + 2NaH(s) \rightarrow Na_2S_2O_4(s) + H_2(g)$$

How many grams of sodium dithionite can be obtained from 100 grams of NaH?

3-15. Potassium superoxide is used in self-contained breathing apparatus. What volume of oxygen (37°C and 1.0 atm) can be obtained from 454 g KO_2?

3-16. Lithium peroxide is used as an oxygen source for self-contained breathing apparatus on space capsules. The relevant reaction is

$$2Li_2O_2(s) + 2CO_2(g) \rightarrow 2Li_2CO_3(s) + O_2(g)$$

What volume of oxygen (at 37°C and 1.0 atm) can be obtained from 454 g Li_2O_2?

3-17. Use the data in Table 3-3 to compute the values of $\Delta\overline{S}_{vap}$ for the alkali metals. Explain briefly why the $\Delta\overline{S}_{vap}$ values are similar. (Hint: Recall Trouton's rule.)

THE ALKALINE EARTH METALS

The Group 2 elements. Top row; beryllium, magnesium, and calcium. Bottom row; strontium and barium.

The Group 2 elements, beryllium, magnesium, calcium, strontium, barium, and radium, are reactive metals with electron configurations of the type [noble gas]ns^2. These elements attain a noble-gas electron configuration by the loss of the two electrons in the outermost *s* orbital.

$$M\{[\text{noble gas}]ns^2\} \rightarrow M^{2+}[\text{noble gas}] + 2e^-$$

They are not as reactive as the Group 1 metals, but they are much too reactive to be found in the free state in nature.

The Group 2 metals also are called the *alkaline earth metals*. Beryllium is a relatively rare element, but occurs as localized

Table 4-1 Major sources and uses of the alkaline earth metals

Metal	*Sources*	*Uses*
beryllium	beryllium aluminum silicates, including *beryl,* $Be_3Al_2Si_6O_{18}$	lightweight alloys (improves corrosion resistance and resistance to fatigue and temperature changes); gyroscopes; nuclear reactors (absorbs neutrons); windows in X-ray tubes
magnesium	*dolomite,* $CaMg(CO_3)_2$; carbonates and silicates; seawater and well brines	alloys for airplanes; flashbulbs; pyrotechnics (white flame); batteries; corrosion protection for metals
calcium	*limestone,* $CaCO_3$; *gypsum,* $CaSO_4 \cdot 2H_2O$; *fluorite,* CaF_2; *apatite,* $Ca_{10}(OH)_2(PO_4)_6$ (major constituent of tooth enamel)	alloys; "getter" (removes gases from electronic tubes); production of chromium and other metals
strontium	*celestite,* $SrSO_4$; *strontianite,* $SrCO_3$	alloys and "getter"
barium	*witherite,* $BaCO_3$; *barite,* $BaSO_4$	lubricant on rotors of anodes in vacuum X-ray tubes; spark-plug alloys
radium	*pitchblende* and *carnotite* ores	skin cancer treatments

surface deposits of the mineral *beryl* (Figure 4-1). Essentially unlimited quantities of magnesium are readily available in seawater, where $Mg^{2+}(aq)$ occurs at a concentration of 0.054 M. Calcium, strontium, and barium rank 5th, 18th, and 19th in abundance in the earth's crust, occurring primarily as carbonates and sulfates (Table 4-1). All isotopes of radium are radioactive, with the longest-lived one (Ra-226) having a half-life of 1600 years.

The chemistry of the Group 2 elements involves primarily the metals and the +2 ions. With few exceptions the reactivity of the Group 2 elements increases from beryllium to barium. As in all the *s*-block and *p*-block groups, the first member of the family differs in several respects from the other members of the family. The anomalous properties of beryllium are attributed to the very small ionic radius of Be^{2+}. The radius of Be^{2+} is similar to that of Al^{3+}, and beryllium(II) has some chemical properties like those of aluminum(III), in keeping with the diagonal relationships found between the first member of a group and the second member of the following group.

The atomic and physical properties of the Group 2 elements are given in Tables 4-2 and 4-3. The periodic trends in the atomic properties of the Group 2 elements are shown clearly in the data in Table 4-2, except for radium, which in some cases appears anomalous. As we go down the group, the ionization

Smithsonian

Figure 4-1 The mineral beryl $Be_3Al_2Si_6O_{18}$, occurs in light-green hexagonal prisms. Beryl is the chief source of beryllium and is used as a gem.

Table 4-2 The atomic properties of the Group 2 elements

Property	*Beryllium*	*Magnesium*	*Calcium*	*Strontium*	*Barium*	*Radium*
chemical symbol	Be	Mg	Ca	Sr	Ba	Ra
atomic number	4	12	20	38	56	88
atomic mass	9.0218	24.305	40.08	87.62	137.33	(223)
number of naturally occurring isotopes	1	3	6	4	7	4
Ground-state electron configuration	$[He]2s^2$	$[Ne]3s^2$	$[Ar]4s^2$	$[Kr]5s^2$	$[Xe]6s^2$	$[Rn]7s^2$
metal radius/pm	110	160	190	210	220	225
ionic radius of M^{2+} ion/pm	31	65	94	110	129	150
sum of first and second ionization energies of M(*g*)/ $kJ \cdot mol^{-1}$	2656	2187	1734	1608	1462	~1480
Pauling electronegativity	1.5	1.2	1.0	1.0	0.9	0.9

energy and the electronegativity decrease, whereas the atomic radii and ionic radii increase. These trends are a direct consequence of the increase in size of the atoms and ions with increase in atomic number.

The molar enthalpies of fusion and vaporization decrease (except for Ra) as we descend the group because of the increase in size of the atoms. The standard reduction voltages, E^0, become more negative as we descend the group, which is opposite to the trend exhibited by the Group 1 metals. The melting points, boiling points, and densities show irregularities in their trends, which are not explained easily.

Table 4-3 The physical properties of the Group 2 elements

Property	*Beryllium*	*Magnesium*	*Calcium*	*Strontium*	*Barium*	*Radium*
melting point/°C	1278	651	845	769	725	~700
boiling point/°C	2970	1107	1487	1384	1740	~1740
density at 20°C/$g \cdot cm^{-3}$	1.85	1.74	1.55	2.54	3.51	6.0
$\Delta \overline{H}_{fus}/kJ \cdot mol^{-1}$	9.8	9.2	9.1	8.2	7.5	8.0
$\Delta \overline{H}_{vap}/kJ \cdot mol^{-1}$	140	95	80	76	65	110
E^0/V at 25°C for $M^{2+}(aq) + 2e^- \rightarrow M(s)$	−1.85	−2.36	−2.87	−2.89	−2.91	−2.92

4-1 THE SMALL SIZE OF Be^{2+} MAKES THE CHEMISTRY OF BERYLLIUM DIFFERENT FROM THAT OF THE OTHER GROUP 2 METALS

The chemistry of beryllium is significantly different from that of the other Group 2 elements, because of the small size of the Be^{2+} ion. All Be(II) compounds involve appreciable covalent bonding and there are no crystalline compounds or solutions involving Be^{2+} as such. The other Group 2 metals have larger sizes and lower ionization energies, making them more electropositive than beryllium. As a consequence, the ionic nature of the compounds of the alkaline earth metals increases down through the group.

Beryllium metal is steel gray, light, very hard, and high-melting. The free element is prepared on a commercial scale by electrolysis of the halides and by reduction of BeF_2 with magnesium.

Beryllium metal is fairly unreactive at room temperature. Hot (400°C) beryllium metal reacts with oxygen to form the oxide, $BeO(s)$, with nitrogen to form the nitride, $Be_3N_2(s)$, and with halogens to form the halides, BeX_2. Some of the more common reactions of beryllium are diagrammed in Figure 4-2. Beryllium is amphoteric. Aqueous solutions of Be(II) salts are acidic, owing to the acid dissociation of $Be(H_2O)_4^{2+}(aq)$:

$$Be(H_2O)_4^{2+}(aq) + H_2O(l) \rightleftharpoons BeOH(H_2O)_3^{+} + H_3O^{+}(aq)$$

In strong base the $Be(H_2O)_4^{2+}(aq)$ ion is converted to the beryllate ion, $Be(OH)_4^{2-}$:

$$Be(H_2O)_4^{2+}(aq) + 4OH^{-}(aq) \rightarrow Be(OH)_4^{2-}(aq) + 4H_2O(l)$$

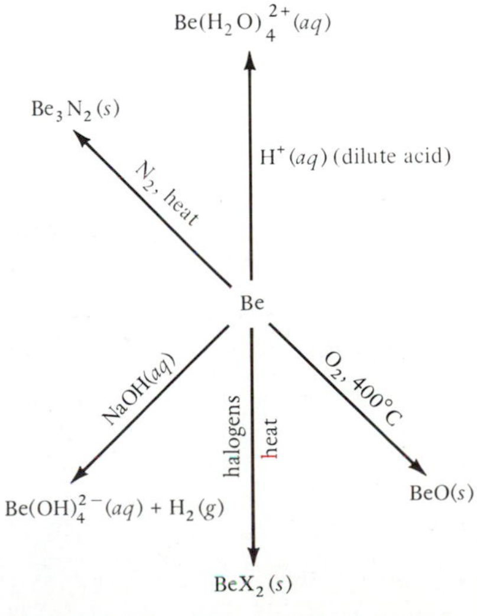

Figure 4-2 Some reactions of beryllium.

4-2 MAGNESIUM, CALCIUM, STRONTIUM, AND BARIUM FORM IONIC COMPOUNDS INVOLVING M^{2+} IONS

The alkaline earth metals are prepared by electrolysis. Magnesium, calcium, strontium, and barium are prepared by high-temperature electrolysis of the molten chloride; for example,

$$CaCl_2(l) \xrightarrow[\text{high T}]{\text{electrolysis}} Ca(l) + Cl_2(g)$$

The metals Mg, Ca, Sr, Ba, and Ra are silvery-white in appearance when freshly cut, but tarnish readily in air to form the metal oxide. The free metals have limited commercial use (Table 4-1). The metals are highly electropositive and readily form M^{2+} ions. The alkaline earth $M^{2+}(aq)$ ions are neutral in aqueous solution.

The alkaline earth metals react rapidly with water, but the rates of these reactions are much lower than those for the alkali metals. Beryllium and magnesium react slowly with water at ordinary temperatures, although hot magnesium reacts violently with water.

Figure 4-3 Magnesium metal burns in oxygen.

The alkaline earth metals burn in oxygen to form the MO oxides, which are ionic solids. Magnesium is used as an incendiary in warfare because of its vigorous reaction with oxygen (Figure 4-3). It burns even more rapidly when sprayed with water and reacts with carbon dioxide via the reaction

$$2Mg(s) + CO_2(g) \rightarrow 2MgO(s) + C(s)$$

Covering burning magnesium with sand slows the combustion, but the molten magnesium reacts with silicon dioxide (the principal component of sand) to form magnesium oxide:

$$2Mg(l) + SiO_2(s) \rightarrow 2MgO(s) + Si(s)$$

Magnesium ribbon is used in flashbulbs. The brilliant flash is produced by the reaction of magnesium with oxygen.

As with the alkali metals, the alkaline earth metals show an increasing tendency to form peroxides with increasing size. Strontium peroxide, SrO_2, is formed at high oxygen pressure, and barium peroxide, BaO_2, forms readily in air at 500°C.

Except for beryllium, the alkaline earth metals react vigorously with dilute acids:

$$Mg(s) + 2HCl(aq) \rightarrow MgCl_2(aq) + H_2(g)$$

Beryllium reacts slowly with dilute acids.

The alkaline earth metals Mg, Ca, Sr, and Ba react with most of the nonmetals to form ionic binary compounds. Their reactions are summarized in Figure 4-4.

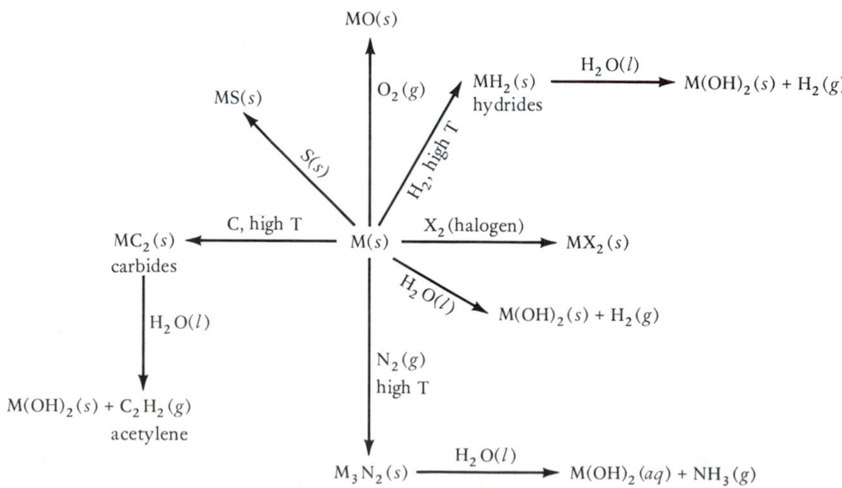

Figure 4-4 Representative reactions of Group 2 metals.

4-3 MAGNESIUM FORMS BOTH IONIC BONDS AND COVALENT BONDS

In addition to forming Mg^{2+} ions, magnesium also exhibits a tendency toward covalent bond formation. In this sense its chemistry differs from calcium, strontium, and barium, but the differences are not as great as those exhibited by beryllium. Covalently bonded organomagnesium halide compounds of the type RMgX (for example, C_2H_5MgBr) are called *Grignard reagents*. They are prepared by the direct reaction of magnesium with an alkyl halide under anhydrous conditions in an electron-donor solvent such as ether. For example,

$$CH_3CH_2Br(\textit{ether}) + Mg(s) \xrightarrow[\text{ether}(l)]{} CH_3CH_2MgBr(\textit{ether})$$

Grignard reagents are used in organic chemistry to synthesize alcohols from carbonyl compounds, which have the general formula

$$\begin{matrix} R \diagdown \\ \quad C{=}O \\ R' \diagup \end{matrix}$$

where R and R′ may be hydrogen atoms or hydrocarbon groups such as $—CH_3$ (methyl) or $—CH_2CH_3$ (ethyl). The synthesis involves a two-step process.

1. Addition of the Grinard reagent to the carbonyl compound in an ether solution. For example,

$$\begin{matrix} H \diagdown \\ \quad C{=}O \\ H \diagup \end{matrix} + CH_3CH_2MgCl \xrightarrow{\text{ether}} CH_3CH_2{—}\overset{\displaystyle H}{\underset{\displaystyle H}{\overset{|}{\underset{|}{C}}}}{—}O{—}Mg{—}Cl$$

2. The adduct is hydrolyzed in an acidic aqueous solution. For example,

$$CH_3CH_2{—}\overset{\displaystyle H}{\underset{\displaystyle H}{\overset{|}{\underset{|}{C}}}}{—}O{—}Mg{—}Cl + HCl(aq) \longrightarrow$$

$$CH_3CH_2{—}\overset{\displaystyle H}{\underset{\displaystyle H}{\overset{|}{\underset{|}{C}}}}{—}OH(aq) + MgCl_2(aq)$$

A wide variety of alcohols can be synthesized by the appropriate choice of Grignard reagent and carbonyl compound. Unlike most organic reactions, these reactions often go to completion.

Figure 4-5 Stalactites and stalagmites are produced when calcium carbonate precipitates from ground water. Shown here is the Powerhouse Cave in West Virginia.

4-4 MANY ALKALINE EARTH METAL COMPOUNDS ARE IMPORTANT COMMERCIALLY

Magnesium sulfate heptahydrate, $MgSO_4 \cdot 7H_2O$, known as Epsom salt, is used as a cathartic, or purgative. The name Epsom comes from the place where the compound was first discovered in 1695, in a natural spring in Epsom, England. Magnesium hydroxide is only slightly soluble in water, and suspensions of it are sold as the antacid Milk of Magnesia.

Calcium is an essential constituent of bones and teeth, limestone (Figure 4-5), plants, and the shells of marine organisms. The Ca^{2+} ion plays a major role in muscle contraction, vision, and nerve excitation. Calcium oxide, or quicklime, is made by heating limestone:

$$CaCO_3(s) \rightarrow CaO(s) + CO_2(g)$$

Calcium oxide is the third ranked industrial chemical; over 35 billion pounds are produced annually in the United States. Large quantities of calcium oxide are used in the steel industry. It is mixed with water to form calcium hydroxide, which is also called *slaked lime:*

$$CaO(s) + H_2O(l) \rightarrow Ca(OH)_2(aq)$$

Slaked lime is used to make cement, mortar, and plaster. Plaster of Paris is $CaSO_4 \cdot \frac{1}{2}H_2O$, which combines with water to form gypsum:

$$\underset{\text{plaster of Paris}}{CaSO_4 \cdot \tfrac{1}{2}H_2O(s)} + \tfrac{3}{2}H_2O(l) \rightarrow \underset{\text{gypsum}}{CaSO_4 \cdot 2H_2O(s)}$$

Asbestos is a calcium magnesium silicate with the approximate composition $CaMg_3(SiO_3)_4$. It can resist very high temperatures, but, because small asbestos fibers are a confirmed carcinogen, it is being phased out as a construction material.

Strontium salts produce a brilliant red flame and are used in signal flares and fireworks (Figure 4-6). The radioactive isotope strontium-90, which is produced in atomic bomb explosions, is a major health hazard because it behaves like calcium and incorporates in bone marrow, causing various cancers.

Some commercially useful compounds of the Group 2 elements are listed in Table 4-4.

Figure 4-6 A red signal flare. The red color arises from light emitted by electronically excited strontium atoms.

Table 4-4 Some important compounds of the Group 2 elements

Compound	*Uses*
beryllium fluoride, $BeF_2(s)$	glass manufacture and nuclear reactors
beryllium oxide, $BeO(s)$	nuclear reactor fuel moderator, electrical insulator
magnesium chloride, $MgCl_2(s)$	fireproofing wood, and disinfectants
magnesium oxide, $MgO(s)$	talcum powder, component of fire bricks; optical instruments
magnesium perchlorate, $Mg(ClO_4)_2(s)$	desiccant
magnesium sulfite, $MgSO_3(s)$	paper pulp manufacture
calcium hydrogen sulfite, $Ca(HSO_3)_2(s)$	germicide, preservative, disinfectant; beer manufacture
calcium carbonate, $CaCO_3(s)$	antacid in wine-making; manufacture of pharmaceuticals
calcium chloride, $CaCl_2(s)$	de-icer on roads, to keep dust down on dirt roads, fire extinguishers
calcium hypochlorite, $Ca(OCl)_2(s)$	bleaching powder, sugar refining, algicide
calcium tartrate, $CaC_4H_4O_6(s)$	fruit and seafood preservative
strontium bromide, $SrBr_2(s)$	sedative and anticonvulsant
strontium hydroxide, $Sr(OH)_2(s)$	sugar refining
strontium nitrate, $Sr(NO_3)_2(s)$	signal flares
strontium sulfide, $SrS(s)$	luminous paints
barium carbonate, $BaCO_3(s)$	rat poison
barium chloride, $BaCl_2(s)$	cardiac stimulant
barium nitrate, $Ba(NO_3)_2(s)$	pyrotechnics (green flame); signal flares
barium selenide, $BaSe(s)$	photocells and semiconductors

TERMS YOU SHOULD KNOW

alkaline earth metals
beryl
Grignard reagent
Epsom salt
slaked lime

QUESTIONS

4-1. Complete and balance the following equations.

(a) $Ca(s) + H_2(g) \xrightarrow{500°C}$

(b) $Mg(s) + N_2(g) \xrightarrow{500°C}$

(c) $Sr(s) + S(s) \xrightarrow{500°C}$

(d) $Ba(s) + O_2(g) \xrightarrow{500°C}$

4-2. Complete and balance the following equations.

(a) $Ca(s) + H_2O(l) \rightarrow$

(b) $Sr_3N_2(s) + H_2O(l) \rightarrow$

(c) $CaC_2(s) + H_2O(l) \rightarrow$

(d) $Ca(s) + C(s) \xrightarrow{500°C}$

4-3. Complete and balance the following equations.

(a) $Be(s) + HCl(aq) \rightarrow$

(b) $Be(s) + NaOH(aq) \rightarrow$

(c) $Be(s) + N_2(g) \xrightarrow{500°C}$

(d) $Be(s) + O_2(g) \xrightarrow{400°C}$

4-4. Burning magnesium, which can occur in automobile fires, should not be attacked with either water or carbon dioxide extinguishers. Why not?

4-5. Unlike the other Group 2 hydroxides, beryllium hydroxide is amphoteric. Write balanced chemical equations for the reaction of $Be(OH)_2$ with $HCl(aq)$ and with $NaOH(aq)$.

4-6. Magnesium hydroxide is only slightly soluble in water, but a suspension of magnesium hydroxide (Milk of Magnesia) in water is used as an antacid.

(a) Write a balanced chemical equation for the neutralization of stomach acid ($HCl(aq)$) by Milk of Magnesia.

(b) Given that stomach acid is about 0.10 M $HCl(aq)$, compute the number of milligrams of $Mg(OH)_2(s)$ required to neutralize 1.0 mL of stomach acid.

4-7. Beryllium is prepared on an industrial scale by the electrolysis of molten $BeCl_2$ or K_2BeF_4 and also by the reduction of BeF_2 with magnesium. Write balanced chemical equations to describe the three processes.

4-8. An old industrial preparation of hydrogen peroxide involves the reaction of oxygen with barium oxide at 500°C to form barium peroxide, followed by the treatment of the peroxide with aqueous acid. Write balanced chemical equations for the process.

4-9. Suggest a method for the preparation of magnesium chloride from magnesium carbonate.

4-10. Suggest a method for the preparation of calcium nitrate from calcium carbonate.

4-11. (a) Use VSEPR theory to predict the structure of beryllium chloride, $BeCl_2$.

(b) Use hybrid orbitals to describe the bonding in $BeCl_2$.

4-12. (a) Use VSEPR theory to predict the shape of the tetrafluoroberyllate(II) ion BeF_4^{2-}.

(b) Use hybrid orbitals to describe the bonding in BeF_4^{2-}.

4-13. The solubilities (in grams per 100 mL of solution) of the alkaline earth hydroxides in water at 20°C are

$Mg(OH)_2$	9×10^{-4}	$Sr(OH)_2$	0.93
$Ca(OH)_2$	0.18	$Ba(OH)_2$	5.8

Calculate the pH of a saturated solution in each case.

4-14. Both $BaCO_3$ and $BaSO_4$ are insoluble in basic solution. In acidic solution, $BaCO_3$ dissolves but $BaSO_4$ does not. Explain.

4-15. Write the formula of the alcohol synthesized from the following combinations:

(a) CH_3CH_2MgBr and $(H)(CH_3)C{=}O$

(b) CH_3CH_2MgBr and $(CH_3)(CH_3)C{=}O$

THE GROUP 3 ELEMENTS

A thermite reaction. As is evident from the figure, thermite reactions are extremely vigorous and highly exothermic. The reaction is driven by the very high stability of the product Al_2O_3.

The Group 3 elements are boron, aluminum, gallium, indium, and thallium. Boron is a semimetal, and the other members of the series are metals, with the metallic character of the elements increasing as we descend the group. The electron configuration of the members of the group is [noble gas]ns^2np^1, and thus the common oxidation states of the Group 3 elements are 0 and +3. The increasing tendency on descending a group to have an oxidation state that is two less than the maximum possible value first appears in Group 3, where In^+ and Tl^+ are significant oxidation states of indium and thallium, respectively.

The chemistry of boron, the first member of the group, differs in many respects from that of the rest of the group.

Table 5-1 The atomic properties of the Group 3 elements

Property	*Boron*	*Aluminum*	*Gallium*	*Indium*	*Thallium*
chemical symbol	B	Al	Ga	In	Tl
atomic number	5	13	31	49	81
atomic mass	10.81	26.98154	69.72	114.82	204.37
number of naturally occurring isotopes	2	1	2	2	2
ground-state electron configuration	$[He]2s^2 2p^1$	$[Ne]3s^2 3p^1$	$[Ar]3d^{10} 4s^2 4p^1$	$[Kr]4d^{10} 5s^2 5p^1$	$[Xe]4f^{14} 5d^{10} 6s^2 6p^1$
metal radius/pm	85	125	130	155	190
ionic radius M^{3+}/pm	20	51	62	81	95 (Tl^+: 1.44)
sum of the first three ionization energies of $M(g)$/kJ · mol^{-1}	6886	5137	5520	5063	5415
Pauling electronegativity	1.9	1.5	1.6	1.7	1.8

Boron, a semimetal, behaves more like the semimetal silicon than like the metal aluminum.

As in Groups 1 and 2, atomic and ionic radii and density increase on descending the group (Tables 5-1 and 5-2). The enthalpies of fusion (except that for Tl) and vaporization, and the boiling point also decrease on descending the group (Table 5-2). These trends are a direct consequence of the increase in size and mass with increase in atomic number. The electronegativities and standard reduction voltages do not show smooth trends, and the observed variations are not explained easily.

Table 5-2 The physical properties of the Group 3 elements

Property	*Boron*	*Aluminum*	*Gallium*	*Indium*	*Thallium*
melting point/°C	2180	660	30	157	304
boiling point/°C	~3650	2467	2250	2070	1457
density at 20°C/g · cm^{-3}	2.35	2.70	5.90	7.30	11.85
$\Delta\overline{H}_{fus}$/kJ · mol^{-1}	23.6	10.5	5.6	3.3	4.3
$\Delta\overline{H}_{vap}$/kJ · mol^{-1}	505	291	270	232	166
E^0/V at 25°C for $M^{3+}(aq) + 3e^- \rightarrow M(s)$	−0.87	−1.66	−0.53	−0.34	+0.72

Table 5-3 Major sources and uses of the Group 3 metals

Metal	*Sources*	*Uses*
boron	kernite, $Na_2B_4O_7 \cdot 4H_2O$ borax, $Na_2B_4O_7 \cdot 10H_2O$ colemanite, $Ca_2B_6O_{11} \cdot 5H_2O$	shield for nuclear radiation, and in instruments used for absorbing and detecting neutrons; hardening agent in alloys
aluminum	bauxite, AlO(OH); clays	in aircraft and rockets, utensils, electrical conductors, photography, explosives, fireworks, paint, building decoration, and telescope mirrors
gallium	trace impurity in bauxite and zinc and copper minerals; by-product in the production of aluminum	semiconductors, LEDs, light-emitting diodes, high-temperature heat-transfer fluid
indium	by-product in lead and zinc production	low-melting alloys in safety devices, sprinkler use, transistor manufacture
thallium	by-product from production of other metals	no significant commercial uses

The major sources and commercial uses of the Group 3 elements are given in Table 5-3.

5-1 THE BONDING IN BORON COMPOUNDS IS COVALENT

Boron is a relatively rare element, but *borax* ($Na_2B_4O_7 \cdot 10H_2O$) was known and used thousands of years ago to glaze pottery. Large deposits of the boron minerals *kernite* ($Na_2B_4O_7 \cdot 4H_2O$) and *borax* are found in certain desert regions of California (Figure 5-1). Boron exists in several allotropic forms with rather complex atomic structures, and which are difficult to obtain in high purity. Boron usually occurs as a brown-black powder.

Boron(III) is always covalently bonded; boron forms no simple cations of the type B^{3+}. For example, the boron trihalides, BX_3, are trigonal planar molecules with X—B—X bond angles of 120° as predicted by VSEPR theory (Chapter 11 of the text).

U.S. Borax

The semimetal boron in crystalline form.

5-2 BX_3 COMPOUNDS ARE LEWIS ACIDS

The boron trihalides are *electron-deficient* compounds whose bonding can be described using sp^2 hybrid orbitals. They have a vacant *p*-orbital perpendicular to the plane of the molecule, and this vacant *p*-orbital can act as an electron acceptor. Thus, the

U.S. Borax

Figure 5-1 Aerial view of a U.S. Borax mine in Boron, California, where massive deposits of borax are located.

boron trihalides are *Lewis acids* and are capable of reacting with electron pair donors, or *Lewis bases;* for example,

$$\underset{\text{Lewis base}}{F^-(aq)} + \underset{\text{Lewis acid}}{BF_3(g)} \rightarrow BF_4^-(aq)$$

$$\underset{\text{Lewis base}}{(CH_3)_3N(aq)} + \underset{\text{Lewis acid}}{BCl_3(g)} \rightarrow (CH_3)_3NBCl_3(aq)$$

The boron trihalides react with water to form boric acid, $B(OH)_3$, and the hydrohalic acid. For example,

$$BCl_3(g) + 3H_2O(l) \rightarrow B(OH)_3(s) + 3H^+(aq) + 3Cl^-(aq)$$

Boric acid is usually made by adding hydrochloric acid or sulfuric acid to borax:

$$Na_2B_4O_7 \cdot 10H_2O(s) + 2HCl(aq) \rightarrow 4B(OH)_3(aq) + 5H_2O(l) + 2NaCl(aq)$$

Boric acid is a moderately soluble monoprotic weak acid in water; its formula is usually written as $B(OH)_3$ rather than

H_3BO_3 or $HBO(OH)_2$, because it acts as a Lewis acid by accepting a hydroxide ion rather than by donating a proton:

$$B(OH)_3(aq) + H_2O(l) \rightleftharpoons B(OH)_4^-(aq) + H^+(aq) \qquad pK_a = 9.3 \text{ at } 25°C$$

Aqueous solutions of boric acid are used in mouth and eye washes.

The principal oxide of boron, $B_2O_3(s)$, is obtained by heating boric acid:

$$2B(OH)_3(s) \rightarrow B_2O_3(s) + 3H_2O(g)$$

Boron trioxide, commonly known as boric oxide, is a colorless, vitreous substance that is extremely difficult to crystallize. It is used to make heat-resistant glassware such as Pyrex, and as a fire-resistant additive for paints.

5-3 BORON HYDRIDES INVOLVE MULTICENTER BONDS

Boron forms a large number of hydrides. The boron hydrides are volatile, spontaneously flammable in air, and easily hydrolyzed. In Chapter 11 of the text we used VSEPR theory to predict that borane, BH_3, is a symmetrical trigonal planar molecule with 120° bond angles, and in Chapter 12 we described the bonding in BH_3 using sp^2 hybrid orbitals. This molecule is not stable, however, and dimerizes to diborane, B_2H_6:

$$2BH_3(g) \rightleftharpoons B_2H_6(g) \qquad K = 10^5 \text{ atm}^{-1} \text{ at } 25°C$$
$$\Delta H^\circ_{rxn} = -145 \text{ kJ}$$

Diborane, which is the simplest boron hydride that can be synthesized in appreciable quantities, can be prepared by reacting sodium borohydride, $NaBH_4$, with sulfuric acid, $H_2SO_4(aq)$:

$$2NaBH_4(aq) + H_2SO_4(aq) \rightleftharpoons B_2H_6(g) + 2H_2(g) + Na_2SO_4(aq)$$

The reaction of B_2H_6 with water is slow enough to enable B_2H_6 to escape from the solution as the gas. Pyrolysis of diborane, in some cases in the presence of H_2, is used to prepare more complex boron hydrides such as B_4H_{10}, B_5H_9, B_6H_{12}, and $B_{10}H_{14}$ and boron-hydride anions such as $B_3H_8^-$ and $B_{12}H_{12}^{2-}$, all of which involve B—H—B bonds (Figure 5-2).

▪ Pyrolysis involves the transformation of one substance into another by heat without oxidation.

The bonding in the boranes cannot be explained by the ideas developed in Chapter 12 of the text. For example, the bonding in diborane is not analogous to the bonding in ethane, C_2H_6 (Figures 12-6 and 12-7 of the text); diborane has two less fewer valence electrons than ethane. The structure of diborane is shown in Figure 5-3. The bonding can be described in terms of sp^3 hybrid orbitals on the boron atoms. Each of the boron-hydrogen bonds at the ends of the molecule involves an sp^3

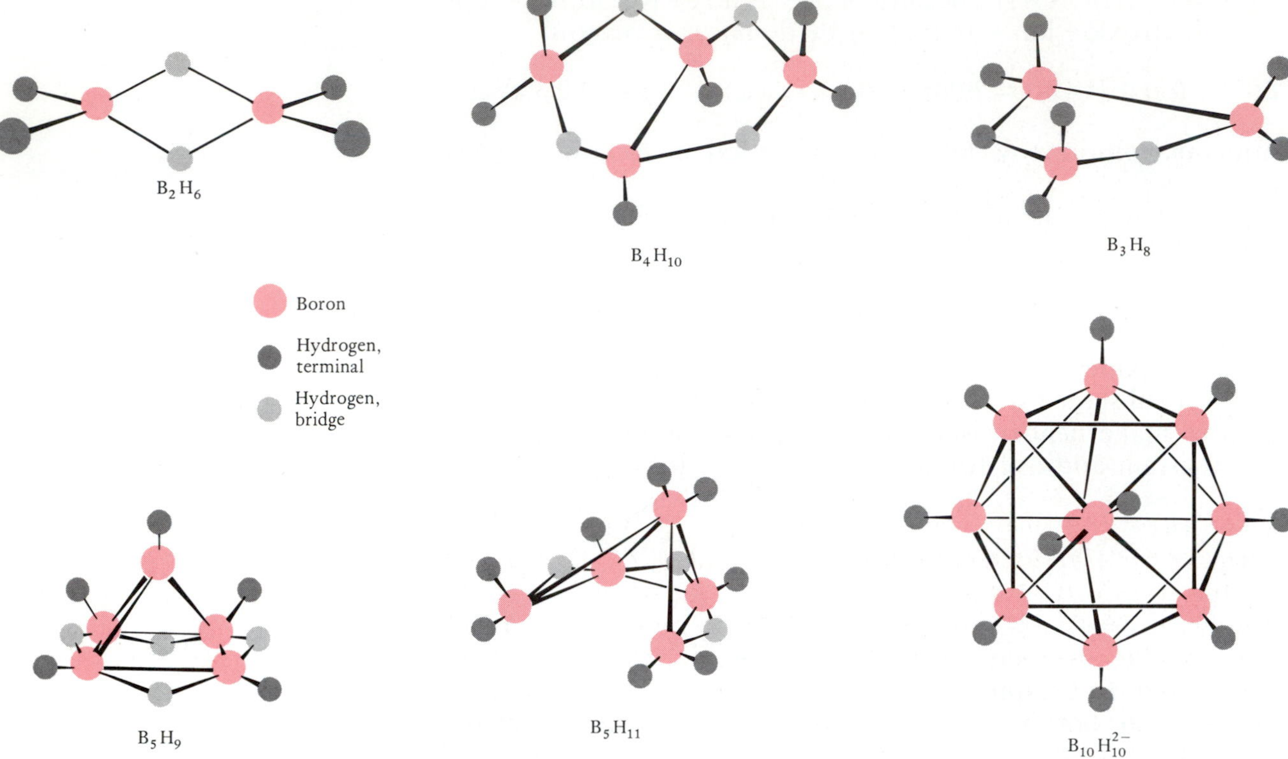

Figure 5-2 Structures of some boron hydrides.

orbital on the boron atom and the 1*s* orbital on the hydrogen atom. The four terminal B—H bonds use eight of the 12 valence electrons. The bonds in the center of the molecule are quite different. Each of the localized bond orbitals spreads over the two boron atoms and a hydrogen atom and can be described as a combination of an sp^3 orbital from each boron atom and the 1*s* orbital from the hydrogen atom. The resulting bond orbitals are

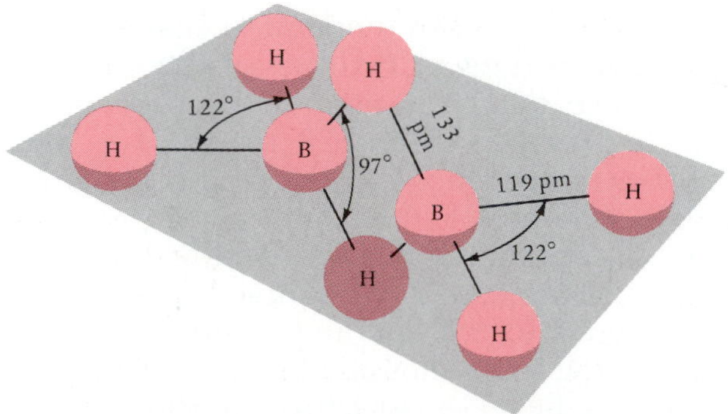

Figure 5-3 The structure of diborane. There are two nonequivalent sets of hydrogen atoms, four lying in a plane with one above and one below the plane.

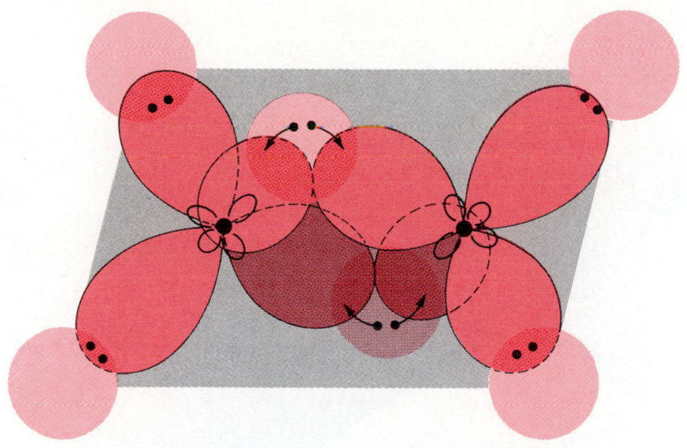

Figure 5-4 An illustration of the bonding orbitals of diborane.

called *three-center bond orbitals;* each one is occupied by two electrons of opposite spin to form a *three-center bond.* Figure 5-4 summarizes the bonding in diborane. The bonding in the higher boranes is more involved than in diborane: Not only are there three-center bonds, but in some there are five-center bonds as well.

5-4 ALUMINUM IS THE MOST ABUNDANT METAL IN THE EARTH'S CRUST

Aluminum is the most abundant metallic element and the third most abundant element in the earth's crust. In addition to its widespread occurrence in silicate minerals, aluminum also is found in enormous deposits of *bauxite,* AlO(OH) (Figure 5-5), which is the chief source of aluminum. The bauxite is refined by the *Bayer process.* The first step is to dissolve the ore in an aqueous sodium hydroxide solution. Bauxite, being amphoteric, dissolves to form sodium aluminate, and some of the sand and silicate rock dissolve to form sodium silicate.

$$2\text{AlO(OH)}(s) + 2\text{NaOH}(aq) + 2\text{H}_2\text{O}(l) \rightarrow 2\text{NaAl(OH)}_4(aq)$$

$$\underset{\text{sand}}{\text{SiO}_2(s)} + 2\text{NaOH}(aq) \rightarrow \text{Na}_2\text{SiO}_3(aq) + \text{H}_2\text{O}(l)$$

The aluminate-silicate solution is cooled and seeded with AlO(OH) or Al_2O_3, which precipitates $Al(OH)_3$ but leaves the silicate in solution. The resultant $Al(OH)_3$ is allowed to react with HF and NaOH in a lead vessel to obtain cryolite, Na_3AlF_6:

$$6\text{HF}(g) + \text{Al(OH)}_3(s) + 3\text{NaOH}(s) \rightarrow \underset{\text{cryolite}}{\text{Na}_3\text{AlF}_6(s)} + 6\text{H}_2\text{O}(g)$$

The cryolite is used to obtain aluminum metal by the *Hall process* (Chapter 21 of the text), where a molten mixture of cryolite,

Tom Carroll/Martin Marietta

Figure 5-5 Bauxite, a reddish-brown ore, is the principal source of aluminum. Bauxite is heated with coke (the black substance shown) to produce aluminum oxide, a white powder. This is further refined through electrolysis to produce aluminum.

together with CaF_2 and NaF, is electrolyzed at 800 to 1000°C. The other Group 3 metals are also obtained by electrolysis of the appropriate molten halide salt, or by electrolysis of aqueous solutions of their salts.

Aluminum is a light, soft metal that resists corrosion by the formation of a tough, adherent protective layer of the oxide, Al_2O_3. Structural alloys of aluminum for aircraft and automobiles contain silicon, copper, magnesium, and other metals to increase the strength and stiffness of the aluminum. Gallium, indium, and thallium are soft, silvery-white metals. Gallium melts at a temperature less than body temperature (Figure 5-6) and has the widest liquid range (2220°C) of any known substance. Indium is soft enough to find use as a metallic O-ring material in metal high-vacuum fittings.

5-5 THE GROUP 3 OXIDES BECOME INCREASINGLY BASIC ON DESCENDING THE GROUP

The Group 3 metals Al, Ga, In, and Tl exhibit two important trends that are also found to varying degrees in Groups 4, 5, 6, and 7. The two trends that are found on descending Group 3 are

1. A decrease in the stability of a higher oxidation state relative to a lower oxidation state. In the case of the Group 3 metals, we find a decrease in the stability of the M(III) state relative to the M(I) state. Although the trivalent state is important for all Group 3 metals, Tl(I) is also an important oxidation state in the chemistry of thallium.

Figure 5-6 Gallium metal has a melting point of 30°C, so a piece of gallium melts when held in the hand (human body temperature is 37°C). Only two elements are liquids at room temperature (20°C). One is the metal mercury, and the other is the nonmetal bromine.

2. An increase in the metallic character for identical oxidation states. This increase in metallic character is illustrated by an increase in the basicity of the oxides.

Aluminum and gallium are amphoteric; they dissolve in both strong aqueous acids and bases:

$$2Al(s) + 6H^+(aq) \rightarrow 2Al^{3+}(aq) + 3H_2(g)$$

$$2Al(s) + 6H_2O(l) + 2OH^-(aq) \rightarrow \underset{\text{aluminate ion}}{2Al(OH)_4^-(aq)} + 3H_2(g)$$

The reaction of aluminum with concentrated aqueous sodium hydroxide is used in the commercial drain cleaner Drāno (Figure 5-7). The heat and gas evolved in the reaction melt the grease and agitate the solid materials blocking the drain, respectively.

The hydroxides and oxides of aluminum and gallium are also amphoteric:

$$Al(OH)_3(s) + OH^-(aq) \rightarrow Al(OH)_4^-(aq)$$

$$Al(OH)_3(s) + 3H^+(aq) \rightarrow Al^{3+}(aq) + 3H_2O(l)$$

Aluminum oxide, which is commonly called alumina, is an extremely stable compound. This stability is evidenced by the ability of aluminum to reduce many metallic oxides to the corresponding metals in the *thermite reaction* (see Frontispiece). For example,

$$2Al(s) + Cr_2O_3(s) \rightarrow Al_2O_3(s) + 2Cr(s) \qquad \Delta H^\circ_{rxn} = -529 \text{ kJ}$$

Figure 5-7 Drāno consists of a mixture of pieces of aluminum and NaOH(*s*). When Drāno is added to water the aluminum reacts with the NaOH(*aq*) to produce hydrogen gas. The overall process is highly exothermic.

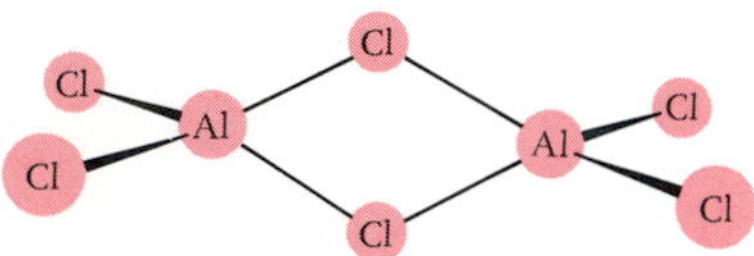

Figure 5-8 The structure of $Al_2Cl_6(g)$. Compare with the structure of diborane (Figure 5-3).

The elements indium (small shiny piece at the left) and thallium (sliced rod showing lustrous metal. The tarnish is due to reaction with air).

Indium and thallium react with aqueous solutions of strong acids, but they are unaffected by strong bases. The oxides and hydroxides of indium and thallium are not amphoteric, but are basic.

5-6 GROUP 3 METALS FORM BOTH IONIC AND COVALENT BONDS

Compounds of the Group 3 metals exhibit both ionic and covalent bonding; however, ionic bonding is somewhat favored. All four metals react with halogens to form compounds with the empirical formula MX_3. The MX_3 fluorides are ionic, whereas the chlorides, bromides, and iodides are low-melting compounds that are dimeric in the vapor state. The halide-bridge structure of the dimer Al_2Cl_6 is shown in Figure 5-8. Note the similarity of the Al_2Cl_6 structure to that of diborane (Figure 5-3).

The $M^{3+}(aq)$ ions of aluminum, gallium, and indium are well-defined cationic species in strongly acidic solutions and undergo acid-dissociation reactions of the type

$$Al(H_2O)_6^{3+}(aq) \rightleftharpoons Al(H_2O)_5OH^-(aq) + H^+(aq)$$

$$K = 1.1 \times 10^{-5}\ \text{M at } 25°\text{C}$$

Aluminum hydroxide occurs as a white, flocculent precipitate that is used to clarify water.

Slow addition of $NaOH(aq)$ to solutions containing $M^{3+}(aq)$ yields the insoluble hydroxides $M(OH)_3(s)$, or hydrated oxides, $M_2O_3 \cdot xH_2O(s)$. Dehydration of the hydroxides or hydrous oxides yields the oxides Al_2O_3 (white), Ga_2O_3 (white), In_2O_3 (yellow), and Tl_2O_3 (brown-black).

The salts $LiAlH_4$ and $LiGaH_4$, which contain the tetrahedral hydride ions MH_4^-, can be prepared from the respective halides and lithium hydride, LiH. For example,

$$4LiH(soln) + AlCl_3(soln) \xrightarrow{\text{ether}} LiAlH_4(soln) + 3LiCl(s)$$

Table 5-4 Some important compounds of the Group 3 elements

Compound	*Uses*
boron trioxide, $B_2O_3(s)$	heat-resistant glassware (Pyrex), fire retardant
boron carbide, $B_4C(s)$	abrasive, wear-resistant tools
boron nitride, $BN(s)$	lubricant, refractory, nose cone windows, cutting tools
aluminum ammonium sulfate, $Al(NH_4)(SO_4)_2(s)$	purification of drinking water, soil acidification, mordant in dyeing
aluminum oxide (alumina), $Al_2O_3(s)$	manufacture of abrasives, refractories, ceramics, spark plugs; artificial gems; absorbing gases and vapors
aluminum borohydride, $Al(BH_4)_3(s)$	reducing agent, rocket fuel component
aluminum hydroxychloride, $AlOHCl_2(s)$	antiperspirant and disinfectant
gallium arsenide, $GaAs(s)$	semiconductors for use in transistors and solar cells
indium oxide, $In_2O_3(s)$	glass manufacture
indium trichloride, $InCl_3(s)$	indium electroplating
thallium(I) oxide, $Tl_2O(s)$	optical glasses and artificial gems
thallium(I) sulfate, $Tl_2SO_4(s)$	rat and ant poison
thallium(I) bromide, $TlBr(s)$, and thallium iodide, $TlI(s)$	crystals for infrared transmitters

These hydrides are useful reducing agents in numerous aprotic solvents, but are violently decomposed by water and occasionally ignite or explode on contact with air.

The M(I) oxidation state of thallium is of major importance in solution. In water at 25°C,

$$Tl^{3+}(aq) + 2e^- \rightleftharpoons Tl^+(aq) \qquad E^0 = +1.25 \text{ V at } 25°\text{C}$$

and thus $Tl^{3+}(aq)$ is about as strong an oxidizing agent as oxygen in 1 M aqueous solutions ($E^0 = 1.23$ V). The thallium(I) ion, Tl^+, is in some respects similar to Ag^+, and in other respects it is similar to K^+ and Rb^+. For example, the nitrate and fluoride salts are water-soluble, whereas the chloride, bromide, chromate, and sulfide salts are insoluble in water. Both TlCl and AgCl are white and darken on exposure to light, whereas the hydroxide $TlOH(aq)$ is a moderately soluble strong base. Thallium(I) compounds are extremely poisonous, and even trace amounts can cause complete loss of body hair.

Some commercially important Group 3 compounds are given in Table 5-4.

TERMS YOU SHOULD KNOW

borax
electron-deficient compound
Lewis acid
Lewis base
pyrolysis
three-center bond orbitals
three-center bonds
bauxite
Bayer process
Hall process
thermite reaction

QUESTIONS

5-1. Complete and balance the following equations.

(a) $Al(s) + Mn_2O_3(s) \xrightarrow[\text{rxn}]{\text{thermite}}$

(b) $Al(C_2H_3O_2)_3(s) + H_2O(l) \rightarrow$

(c) $Al(NO_3)_3(aq) + NH_3(aq) \rightarrow$

(d) $Ga(OH)_3(s) + KOH(aq) \rightarrow$

5-2. Do the acidities of the oxides of the Group 3 oxides increase or decrease upon descending the group?

5-3. Complete and balance the following equations:

(a) $Ga(s) + HCl(aq)$

(b) $Ga(s) + NaOH(aq)$

5-4. Write chemical equations describing the amphoteric nature of $Al(OH)_3(s)$ and $Ga(OH)_3(s)$.

5-5. The formula for boric acid is often written as $B(OH)_3$ rather than H_3BO_3. A good reason for doing this is because boric acid acts not as a Brønsted-Lowry acid but as a Lewis acid. Write a chemical equation describing the Lewis acid property of an aqueous solution of boric acid.

5-6. Write a balanced chemical equation for the formation of $B_{10}H_{14}$ from B_2H_6.

5-7. Diborane is often prepared by the reaction of boron trifluoride with sodium borohydride, with sodium tetrafluoroborate(III) as the other product. Write a balanced chemical equation for this reaction.

5-8. How many valence electrons are there in (a) $B_3H_8^-$; (b) $B_{10}H_{10}^{2-}$?

5-9. Use VSEPR theory to predict the structures of the following species.

(a) AlF_6^{3-} (b) $Al(OH)_4^-$ (c) $AlOF$

5-10. Use VSEPR theory to predict the structures of the following species.

(a) $GaCl_3$ (b) GaF_2^+ (c) $GaBr_4^-$

5-11. Use VSEPR theory to predict the shape of the tetrafluoroborate(III) ion, BF_4^-. Describe the bonding in BF_4^- using hybrid orbitals.

5-12. The structure of tetraborane, B_4H_{10}, is shown in Figure 5-2. Describe the bonding in terms of hybrid orbitals.

5-13. Describe the bonding in $B_3H_8^-$ in terms of hybrid orbitals (see Figure 5-2).

5-14. A mixture of aluminum sulfate and sodium carbonate can be used to clarify water. Write chemical equations to describe this process.

5-15. Use the following data to compute the equilibrium constant at 25°C for the reaction

$$2Tl^+(aq) + O_2(g) + 4H^+(aq) \rightarrow 2Tl^{3+}(aq) + 2H_2O(l)$$

	$E°$/V at 25°C
$Tl^{3+}(aq) + 2e^- \rightleftharpoons Tl^+(aq)$	1.25 V
$O_2(g) + 4H^+(aq) + 4e^- \rightleftharpoons 2H_2O(l)$	1.23 V

5-16. In a thermite reaction involving aluminum and Fe_2O_3, so much heat is evolved that the resulting iron is molten. Given the following data, determine if enough heat is evolved in an $Al–Cr_2O_3$ thermite reaction to melt the chromium. Assume (a) that all the heat evolved is absorbed by the products and (b) a 50 percent heat loss.

	m.p./°C	$\overline{C}_p$/J · mol^{-1} · K^{-1}	$\Delta\overline{H}_{fus}$/J · mol^{-1}
Al_2O_3	2054	79.0	—
Cr	1857	23.4	20.5

5-17. The pK_a values for the Group 3 $M^{3+}(aq)$ ions are as follows:

Ion	pK_a at 25°C
$Al^{3+}(aq)$	4.96
$Ga^{3+}(aq)$	2.60
$In^{3+}(aq)$	2.66
$Tl^{3+}(aq)$	1.15

Compute the pH of a 0.10 M aqueous solution of each of these ions.

5-18. The K_{sp} for $Al(OH)_3(s)$ in water at 25°C is 1.3×10^{-33} M^4. Sodium hydroxide is added slowly to 0.10 M $Al(NO_3)_3(aq)$. Compute the pH of the solution at which $Al(OH)_3(s)$ begins to precipitate.

THE GROUP 4 ELEMENTS

The Group 4 elements. From left to right: Carbon (as graphite), silicon, germanium, tin, lead.

The Group 4 elements, carbon, silicon, germanium, tin, and lead, provide the best example of a group in which the first member has properties different from the rest of the group. Carbon is decidedly nonmetallic and differs markedly from the rest of the members of the group, which becomes increasingly metallic with increasing atomic number. The properties of the remaining members vary more smoothly from silicon to lead: Silicon and germanium are semimetals; and tin and lead are metals. As in Group 3, there is a tendency for the elements with higher atomic number to exhibit an oxidation state of two less than the maximum of +4. The common oxidation state of silicon and germanium is +4, but lead and, to some extent, tin have a common oxidation state of +2.

Westmoreland Coal Co.

In a West Virginia coal mine this 900 horsepower miner cuts through coal and surrounding rock. Its dual cutting heads have replaceable drill tips composed of silicon carbide.

Carbon is widely distributed in nature both as the free element and in compounds. The great majority of carbon occurs in coal, petroleum, limestone ($CaCO_3$), dolomite ($MgCa(CO_3)_2$), and a few other deposits. Carbon is also a principal element in all living matter, and the study of its compounds forms the vast field of organic chemistry.

Silicon constitutes 28 percent of the mass of the earth's mantle and is the second most abundant element in the mantle, being exceeded only by oxygen. Silicon does not occur as the free element in nature; it occurs primarily as the oxide and in numerous silicates. Most sands are essentially silicon dioxide, and many rocks and minerals are silicate materials. Germanium and tin rank in the range fortieth to fiftieth in elemental abundance. The presence of small amounts of germanium in coal deposits serves as a commercial source of that element. The most important source of tin is the mineral *cassiterite,* SnO_2, from which tin is obtained by reduction with coke. Lead is the most abundant of the heavy metals, its most important ore being *galena,* PbS. The principal sources and commercial uses of the Group 4 elements are given in Table 6-1. Tables 6-2 and 6-3 list the atomic properties and the physical properties of the Group 4 elements.

6-1 DIAMOND AND GRAPHITE ARE ALLOTROPIC FORMS OF CARBON

Solid carbon displays the two important allotropic forms, diamond and graphite (Figure 6-1). Diamond has an extended, covalently bonded tetrahedral structure. Each carbon atom lies

▪ Recall that allotropes are two or more forms of an element that differ in their physical or chemical properties.

Table 6-1 Sources and uses of the Group 4 elements

Element	*Principal sources*	*Uses*
carbon	coal and petroleum	fuels; production of iron, furnace linings, electrodes (coke); lubricant, fibers, pencils, airframe structures (graphite); decolorizer, air purification, catalyst (activated charcoal); rubber and printing inks (carbon black); drill bits, abrasives (diamond)
silicon	*quartzite* or sand (SiO_2)	steel alloys, silicones, semiconductor in integrated circuits, rectifiers, transistors, solar batteries
germanium	coal ash; by-product of zinc refining	solid-state electronic devices, alloying agent, phosphor, infrared optics
tin	*cassiterite* (SnO_2)	food packaging, tin plate, pewter, bronze, soft solder
lead	*galena* (PbS), *anglesite* ($PbSO_4$), *cerussite* ($PbCO_3$)	storage batteries, solder and low-melting alloys, type metals, ballast, lead shot, cable covering

at the center of a tetrahedron formed by four other carbon atoms (Figure 6-2). The C—C bond distance is 154 pm, which is the same as the C—C bond distance in ethane. The diamond crystal is, in effect, a gigantic molecule. The hardness of diamond is due to the fact that each carbon atom is attached by strong covalent bonds to four other carbon atoms, and thus many covalent bonds must be broken in order to cleave a

Table 6-2 Atomic properties of the Group 4 elements

Property	*C*	*Si*	*Ge*	*Sn*	*Pb*
atomic number	6	14	32	50	82
atomic mass/amu	12.011	28.0855	72.59	118.69	207.2
number of naturally occurring isotopes	3	3	5	10	4
ground-state electron configuration	$[He]2s^22p^2$	$[Ne]3s^23p^2$	$[Ar]3d^{10}4s^24p^2$	$[Kr]4d^{10}5s^25p^2$	$[Xe]4f^{14}5d^{10}6s^26p^2$
atomic radius/pm	70	110	125	145	180
ionization energy/$MJ \cdot mol^{-1}$					
first	1.09	0.786	0.761	0.708	0.715
second	2.35	1.57	1.53	1.41	1.45
third	4.62	3.23	3.30	2.94	3.08
fourth	6.22	4.36	4.41	3.93	4.08
Pauling electronegativity	2.5	1.8	1.8	1.8	1.8

General Electric

Figure 6-1 Two allotropes of carbon are graphite and diamond.

diamond. Graphite has the unusual layered structure shown in Figure 6-3. The C—C bond distance within a layer is 139 pm, which is close to the C—C bond distance in benzene. The distance between layers is about 340 pm. The bonding within a layer is strong, but the interaction between layers is weak. Therefore, the layers easily slip past each other. This is the molecular basis of the lubricating action of graphite. The "lead" of lead pencils is actually graphite. Layers of the graphite rub off from the pencil onto the paper. Graphite is the stable form at ordinary temperatures and pressures, and diamond is the stable form at high pressures (Figure 13-43 in the text).

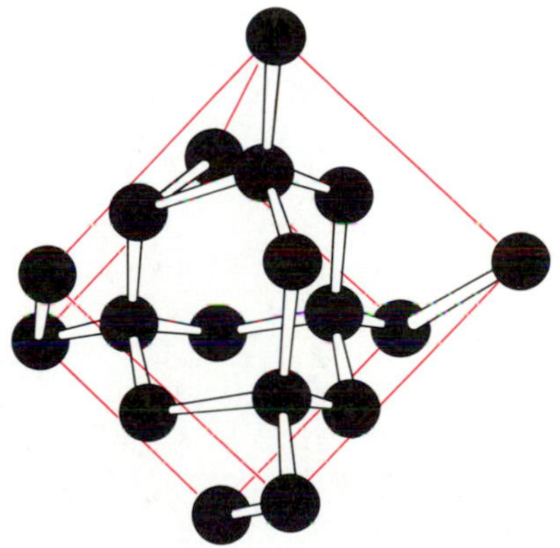

Figure 6-2 The crystalline structure of diamond. Each carbon atom is covalently bonded to four other carbon atoms, forming a tetrahedral network. A diamond crystal is essentially one gigantic molecule.

Table 6-3 Some physical properties of the Group 4 elements

Property	*C*	*Si*	*Ge*	*Sn*	*Pb*
melting point/°C		1410	940	232	328
boiling point/°C	sublimes at ~3900	~3000	2850	2620	1750
$\Delta\overline{H}_{fus}$/kJ · mol^{-1}	105	50.2	36.8	6.99	4.77
$\Delta\overline{H}_{vap}$/kJ · mol^{-1}		359	328	296	179
density at 20°C/g · cm^{-3}	2.27 (graphite)	2.33	5.32	7.28 (white)	11.34

Figure 6-3 The layered structure of graphite; each layer resembles a network of benzene rings joined together. The bonding within a layer is covalent and strong. The interaction between layers, however, is due only to London forces and so is relatively weak. Consequently, the layers easily slip past each other, giving graphite its slippery feel. The density of graphite is 2.2 g · cm^{-3}, which is lower than that of diamond (3.5 g · cm^{-3}), reflecting the more open structure of graphite.

Figure 6-4 The reaction of calcium carbide with water yields acetylene gas and calcium hydroxide. The acetylene gas burns in air and is used to provide light in lamps on hats used by spelunkers.

6-2 CARBON RANKS SECOND AMONG THE ELEMENTS IN THE NUMBER OF COMPOUNDS FORMED

Carbon forms more compounds than any other element except hydrogen. Most of these compounds are classified as organic compounds, which are discussed in Chapter 25 of the text. Although the classification of compounds into inorganic compounds and organic compounds is artificial, we shall discuss a few of the important "inorganic" compounds of carbon here. Binary compounds in which carbon is combined with less electronegative elements are called carbides. For example, aluminum carbide, Al_4C_3, is made by heating aluminum powder or aluminum oxide with coke in an electric furnace. Aluminum carbide is a yellow crystalline substance with a melting point above 2000°C. It hydrolyzes to produce methane according to

$$Al_4C_3(s) + 12H_2O(l) \rightarrow 4Al(OH)_3(s) + 3CH_4(g)$$

This reaction is used to generate methane in certain metallurgical applications. One of the most important carbides is calcium carbide. It is produced industrially by the reaction of lime (CaO) and coke:

$$CaO(s) + 3C(s) \xrightarrow{2000°C} CaC_2(s) + CO(g)$$

Calcium carbide is a gray-black, hard solid with a melting point over 2000°C. In contrast to aluminum carbide, calcium carbide produces acetylene when it hydrolyzes (Figure 6-4):

$$CaC_2(s) + 2H_2O(l) \rightarrow C_2H_2(g) + Ca(OH)_2(s)$$

Carbon tetrabromide is a colorless solid, carbon tetrachloride a clear liquid, and carbon tetraiodide a red crystalline solid.

At one time, this reaction represented one of the major sources of acetylene for the chemical industry and for oxyacetylene welding. One other industrially important carbide is silicon carbide, SiC, known also as *carborundum.* Corborundum is one of the hardest known materials and is used as an abrasive for cutting metals and polishing glass. The structure of carborundum is similar to the cubic crystal structure of diamond.

Carbon forms binary compounds with the halogens. Carbon tetrafluoride, CF_4, or tetrafluoromethane, is a colorless, odorless gas that is chemically unreactive. It can be prepared by the direct combination of carbon and fluorine, and also via the reaction

$$SiC(s) + 4F_2(g) \rightarrow SiF_4(g) + CF_4(g)$$

The mixture of gases produced may be separated by passing them through an alkaline solution. The CF_4 does not react, but the SiF_4 hydrolyzes according to

$$SiF_4(g) + 4H_2O(l) \rightarrow Si(OH)_4(s) + 4HF(aq)$$

The hydrolysis apparently involves the attack on SiF_4 by H_2O, which can occur via the $3d$ orbitals on silicon. Carbon tetrafluoride has no low-lying d orbitals and thus is unreactive. This difference in reactivity between CF_4 and SiF_4 is typical of carbon and silicon compounds.

Carbon tetrachloride is a colorless liquid with a sweetish, characteristic odor. It was used extensively as a solvent and dry-cleaning agent, but its use has declined because of its toxicity. Carbon tetrabromide is a pale-yellow to colorless solid that is markedly less stable than CF_4 or CCl_4. This trend continues with carbon tetraiodide, a bright-red crystalline substance with an odor similar to that of iodine. It readily decomposes under heat according to

$$2CI_4(s) \rightarrow 2I_2(s) + C_2I_4(s)$$

The decreasing stability of the carbon tetrahalides with increasing atomic mass is illustrated nicely by the C—X molar bond enthalpies:

Bond	*Bond enthalpy*/$kJ \cdot mol^{-1}$
C—F	439
C—Cl	331
C—Br	276
C—I	213

6-3 CARBON FORMS A NUMBER OF IMPORTANT INORGANIC COMPOUNDS

Carbon has a number of oxides, but only two of them are particularly stable. When carbon is burned in a limited amount of oxygen, carbon monoxide predominates. When an excess of oxygen is used, carbon dioxide results. Carbon monoxide is an odorless, colorless, tasteless gas that burns in oxygen to produce carbon dioxide. It is highly poisonous owing to the fact that it binds to hemoglobin much more strongly than does oxygen. It is used as a fuel and as a reducing agent in metallurgy. Carbon dioxide is an odorless, colorless gas with a faintly acidic taste. When CO_2 is dissolved in water, a small amount of it is converted to carbonic acid, H_2CO_3. At 25°C, we have for the equilibrium constant of the reaction

$$CO_2(aq) + H_2O(l) \rightarrow H_2CO_3(aq) \qquad K = 1.7 \times 10^{-3}$$

Small amounts of carbon dioxide can be produced by the reaction of carbonates with acids. Over 50 percent of the carbon dioxide produced industrially is used as a refrigerant, either as a liquid or as a solid (Dry Ice), and about 25 percent is used to carbonate soft drinks. The phase diagram of carbon dioxide is discussed in Section 13-11 of the text.

Carbon forms several sulfides, but only one of them, carbon disulfide, CS_2, is stable. Carbon disulfide is a colorless, poisonous, flammable liquid. The purified liquid is said to have a sweet, pleasing odor, but the commonly occurring commercial and reagent grades have an extremely disagreeable odor due to organic impurities. Large quantities of CS_2 are used in the manufacture of rayon, carbon tetrachloride, and cellophane, and as a solvent for a number of substances.

Carbon also forms several important nitrogen-containing compounds. Hydrogen cyanide, HCN, is a colorless, extremely poisonous gas that dissolves in water to form the very weak acid, hydrocyanic acid ($pK_a = 9.32$ at 25°C). Salts of hydrocyanic acid, called cyanides, are prepared by direct neutralization. Sodium cyanide, NaCN, is used in the extraction of gold and silver from their ores (Chapter 11), and in the electroplating industry.

6-4 SILICON IS A SEMIMETAL

Silicon, the most important industrial semimetal, has a gray, metallic luster. Its major use is in the manufacture of transistors. Elemental silicon is made by the high-temperature reduction of silicon dioxide (the major constituent of numerous sands) with carbon:

■ Amorphous silicon is a solid form of silicon that does not have a crystalline structure. It is used in the fabrication of some solar-energy devices, because of its relatively low cost.

$$SiO_2(l) + C(s) \xrightarrow{3000°C} Si(l) + CO_2(g)$$

The 98 percent pure silicon prepared by this reaction must be further purified before it can be used to make transistors. It is converted to the liquid silicon tetrachloride by reaction with chlorine:

$$Si(s) + 2Cl_2(g) \rightarrow SiCl_4(l)$$

The silicon tetrachloride is further purified by repeated distillation and then converted to silicon by reaction with magnesium:

$$SiCl_4(g) + 2Mg(s) \rightarrow 2MgCl_2(s) + Si(l)$$

Philips

Ultrapure (99.9999 percent) silicon is produced in the form of a cylinder and is then sliced into thin wafers for semiconductor manufacture.

Lawrence Livermore Lab

A germanium crystal is being produced by the zone-refining method. This crystal of ultrapure germanium was grown from a melt.

The resulting silicon is purified still further by a special method of recrystallization called *zone refining*. In this process, solid silicon is packed in a tube that is mounted in a vertical position (Figure 6-5) with an electric heating loop around the base of the tube. The solid near the heating loop is melted by passing a current through the loop, and the tube is then lowered very slowly through the loop. As the melted solid cools slowly in the region of the tube below the heating loop, pure crystals separate out, leaving most of the impurities behind in the moving molten zone. The process can be repeated as often as necessary to achieve the desired purity of the recrystallized solid. Purities up to 99.9999 percent are possible with zone refining.

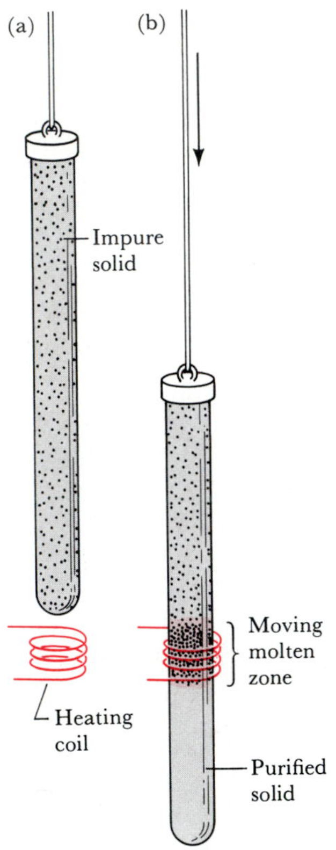

Figure 6-5 Zone refining. An impure solid is packed tightly in a glass tube, and the tube is lowered slowly through a heating coil that melts the solid. Pure solid crystallizes out from the bottom of the melted zone, and the impurities concentrate in moving molten zone.

6-5 THERE ARE TWO TYPES OF SEMICONDUCTORS, *n*-TYPE AND *p*-TYPE

In a crystal, there are two sets of energy levels because of the combination of the valence orbitals of all the atoms. These two sets of energy levels are analogous to the bonding and antibonding orbitals that occur when orbitals from just two atoms are combined (Section 12-10 of the text). The lower set of energy levels is called the *valence band* and is occupied by the valence electrons of the atoms. The higher set is called the *conduction band.* Electrons in the conduction band can move readily throughout the crystal (Figure 6-6).

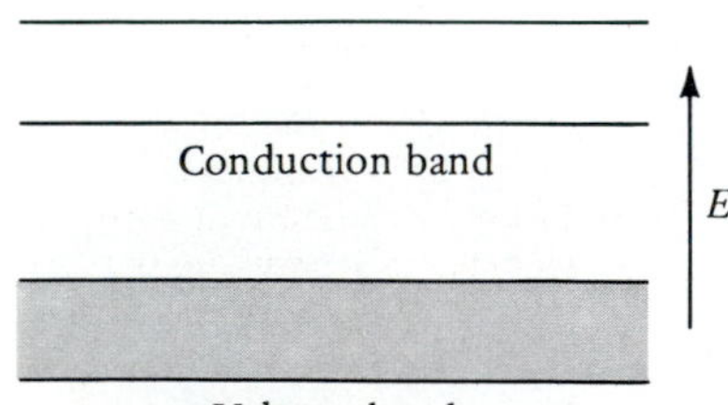

Figure 6-6 When the atoms of a crystal are brought together to form the crystal lattice, the valence orbitals of the atoms combine to form two sets of energy levels, called the valence band and the conduction band.

An electric current is carried in a solid by the electrons in the conduction band, which are called the conduction electrons. In an insulator (such as a nonmetal), there are essentially no electrons in the conduction band because the energies there are much higher than the energies in the valence band. Metals are excellent electrical conductors because there is no energy gap between the conduction band and the valence band. The valence electrons in a metal are conduction electrons. In a semiconductor the energy separation between the conduction band and the valence band is comparable to thermal energies, and thus some of the valence electrons can be thermally excited into the conduction band. Thus a semiconductor has electrical properties intermediate between those of metals and insulators. Figure 6-7 illustrates the difference between a metal, an insulator, and a semiconductor.

An insulator like silicon can be converted to a semiconductor by addition of selected impurity atoms. For example, an *n*-type (*n* for negative) silicon semiconductor is produced when trace amounts of atoms with five valence electrons, such as phosphorus or antimony, are added to silicon, which has four valence electrons (Figure 6-8a). The excess valence electrons on the impurity atoms, which substitute for some of the silicon atoms in the crystal, become the current carriers in the crystal (Figure 6-8b). A *p*-type (*p* for positive) semiconductor is produced when trace amounts of atoms with three valence electrons, such as boron or indium, are added to silicon. The

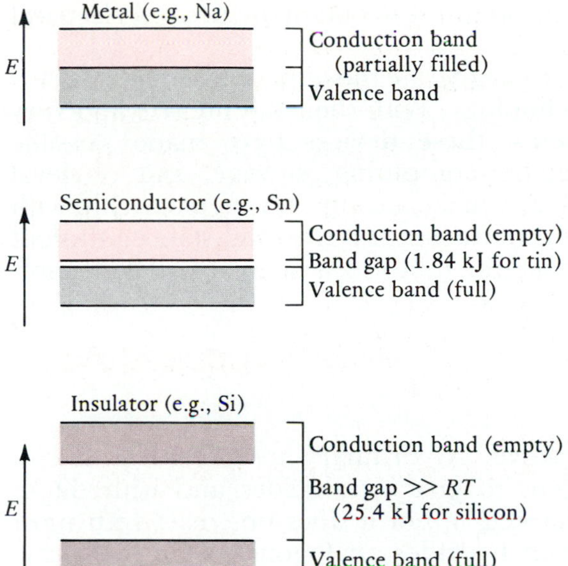

Figure 6-7 A comparison of the energy separations between the valence bands (bonding electron energy levels) and conduction bands (accessible energy levels for mobile electrons) of metals, semiconductors, and insulators.

deficiency of valence electrons on the impurity atoms functions as "holes" by means of which electrons can "hop" through the silicon crystal (Figure 6-8c). Because impurity atoms have a major effect on the electrical properties of semiconductors, it is necessary to use extremely pure ($\geqslant$99.9999 percent) silicon and to add precise amounts of impurities of carefully controlled

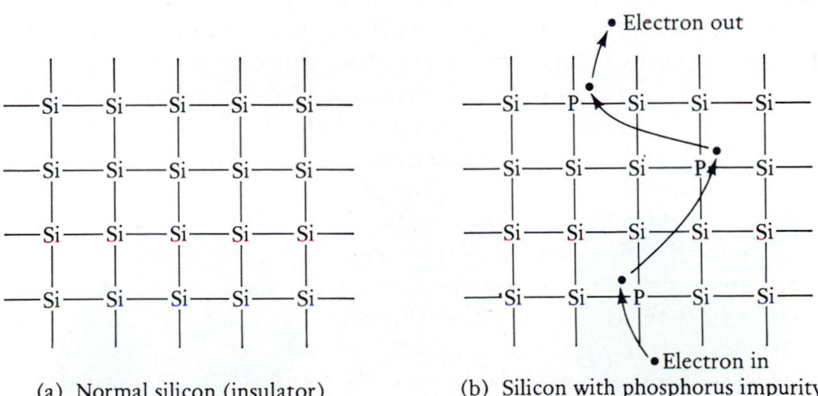

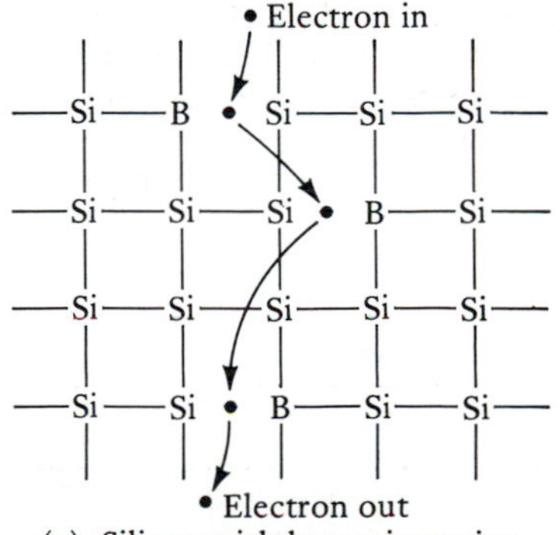

Figure 6-8 Comparison of normal, *n*-type, and *p*-type silicon. (a) Silicon has four valence electrons, and each silicon atom forms four 2-electron bonds to other silicon atoms. (b) Phosphorus has five valence electrons, and thus when a phosphorus atom substitutes for an silicon atom in a silicon crystal, there is an unused valence electron on each phosphorus atom that can become a conduction electron. (c) Boron has only three valence electrons, and thus when a boron atom substitutes for a silicon atom in a silicon crystal, there results an electron vacancy (a "hole"). Electrons from the silicon valence bond can move through the crystal by hopping from one vacancy site to another.

composition to the crystal in order to obtain the desired electrical properties.

It would be difficult to exaggerate the impact of semiconductor devices on modern technology. With their minute size and very low power requirements, these devices have made possible computers with incredible computing, storage, and retrieval capabilities. It is possible to make a computer memory chip with over a million transistors (a transistor is the solid-state equivalent of the now-obsolete vacuum tube) in a space of only 1 mm^2.

▪ The South Bay area of San Francisco is called the Silicon Valley and the area around Austin, Texas, is called the Silicon Prairie because of the large number of companies that produce semiconductors, transistors, and computer chips.

Philips

Computer chips in a contact lens. Each chip may consist of over a million transistors.

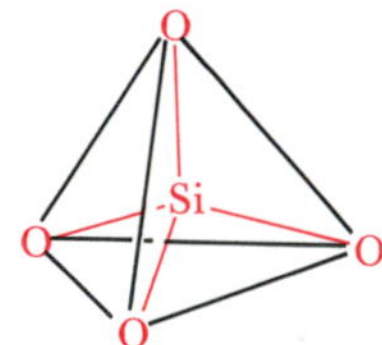

Figure 6-9 The orthosilicate ion, SiO_4^{4-}.

6-6 THE MOST COMMON AND IMPORTANT COMPOUNDS OF SILICON INVOLVE OXYGEN

Silicon is fairly unreactive. At ordinary temperatures silicon reacts with the halogens to give tetrahalides and with dilute alkalis to give silicates (SiO_4^{4-}), but it does not react with most acids. Silanes, the silicon hydrides analogous to the hydrocarbons, are much less stable than the corresponding hydrocarbons. Monosilane, SiH_4, and disilane, Si_2H_6, are thermodynamically stable with respect to the elements at room temperature, but the higher silanes decompose spontaneously at room temperature. There are no silicon analogs of ethylene or acetylene, or of unsaturated hydrocarbons in general.

Silicates occur in numerous minerals and in asbestos, mica, and clays. Cement, bricks, tiles, porcelains, glass, and pottery are all made from silicates. All silicates involve silicon-oxygen single bonds, of which there are two types. Terminal —Si—O bonds involve oxygen bonded to silicon and no other atoms, and bridging —Si—O—Si— bonds involve oxygen linking two silicon atoms.

Figure 6-10 The minerals enstatite (right), willemite (rear), spodumene (front), and zircon (left).

The simplest silicate anion is the tetrahedral *orthosilicate* ion, SiO_4^{4-} (Figure 6-9). The SiO_4^{4-} ion is found in the minerals *zircon*, $ZrSiO_4$, and *willemite*, Zn_2SiO_4, and also in sodium silicate, which, when dissolved in water, is called *water glass*, $Na_4SiO_4(aq)$.

The minerals *enstatite*, $MgSiO_3$, and *spondumene*, $LiAl(SiO_3)_2$ (Figure 6-10), are silicates that contain long, straight-chain silicate polyanions involving the SiO_3^{2-} chain unit:

$$\cdots\!-\!O\!-\!\underset{\underset{O^{\ominus}}{|}}{\overset{\overset{O^{\ominus}}{|}}{Si}}\!-\!O\!-\!\underset{\underset{O^{\ominus}}{|}}{\overset{\overset{O^{\ominus}}{|}}{Si}}\!-\!O\!-\!\underset{\underset{O^{\ominus}}{|}}{\overset{\overset{O^{\ominus}}{|}}{Si}}\!-\!O\!-\!\underset{\underset{O^{\ominus}}{|}}{\overset{\overset{O^{\ominus}}{|}}{Si}}\!-\!O\!-\!\underset{\underset{O^{\ominus}}{|}}{\overset{\overset{O^{\ominus}}{|}}{Si}}\!-\!\cdots$$

$$\underbrace{\quad}_{SiO_3^{2-}}\quad\underbrace{\quad}_{SiO_3^{2-}}\quad\underbrace{\quad}_{SiO_3^{2-}}\quad\underbrace{\quad}_{SiO_3^{2-}}\quad\underbrace{\quad}_{SiO_3^{2-}}$$

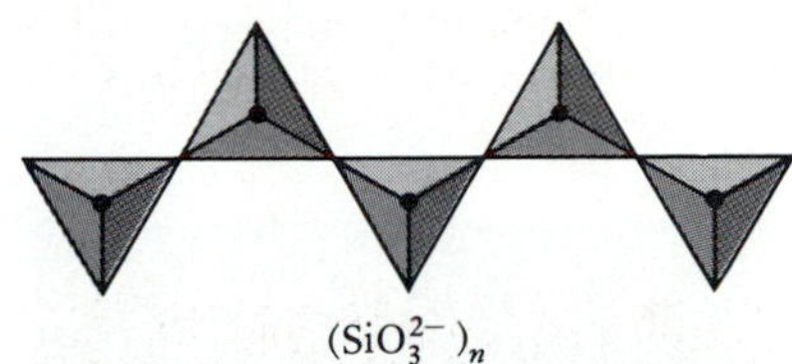

Figure 6-11 Tetrahedral SiO_4^{4-} units are linked together through oxygen atoms that are shared by tetrahedra to form straight-chain silicate polyanions.

Structures that result from joining many smaller units together are called *polymers*. The straight-chain silicate anions shown in the preceding structure are called silicate polyanions because they result from joining together many silicate anions. The mineral *beryl*, $Be_3Al_2Si_6O_{18}$, contains the cyclic polysilicate anion $Si_6O_{18}^{12-}$ (Figure 6-12a). These cyclic polysilicate anions can themselves be joined together to form polymeric, cyclic

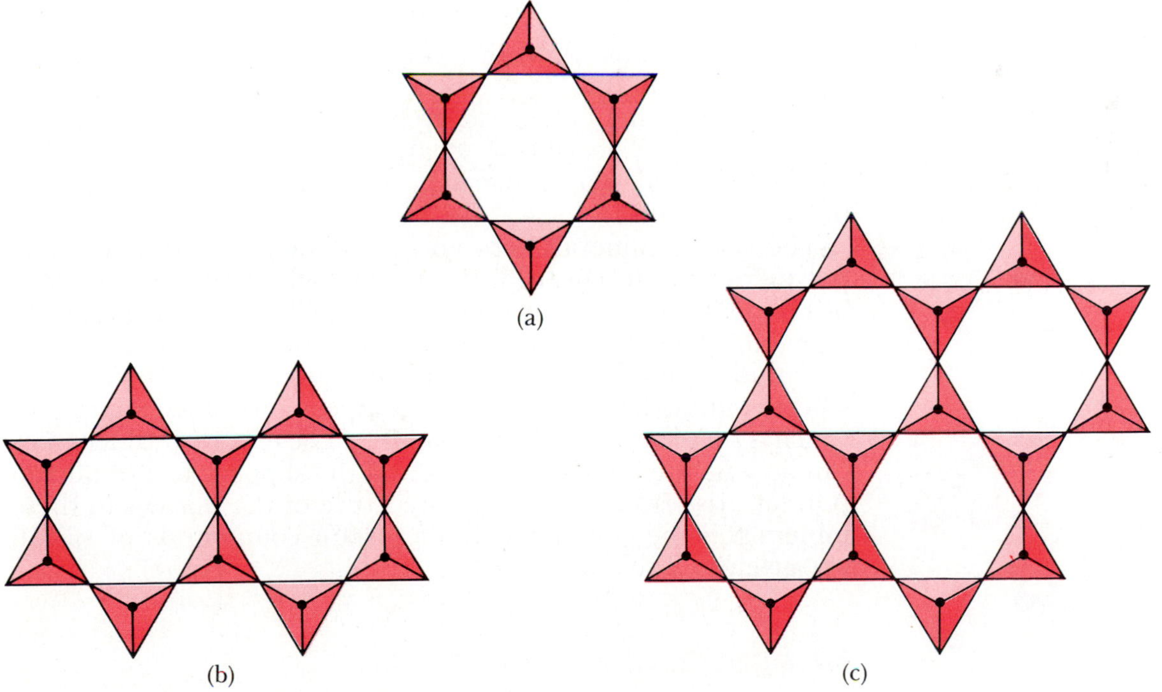

Figure 6-12 (a) The cyclic polysilicate ion $Si_6O_{18}^{12-}$, which occurs in the mineral beryl. Six SiO_4^{4-} tetrahedral units are joined in a ring with the tetrahedra linked by shared oxygen atoms. (b) The cyclic polysilicate ion $Si_6O_{18}^{12-}$ can form a polymeric cyclic network like that shown here. The composition of the cyclic network is $(Si_4O_{11}^{6-})_n$. Asbestos has this structure. (c) Structure of polysilicate sheets composed of $(Si_2O_5^{2-})_n$ subunits. Mica has this structure.

Figure 6-13 Asbestos. The fibrous character of this mineral is a direct consequence of its $(Si_4O_{11}^{6-})_n$ polymeric chains.

Figure 6-14 The ease with which the mineral mica can be separated into thin sheets is a direct consequence of the existence of polymeric silicate sheets with the composition $(Si_2O_5^{2-})_n$.

polysilicate anions with the composition $(Si_4O_{11}^{6-})_n$ and the structure shown in Figure 6-12(b). The best example of a mineral containing polymeric, cyclic polysilicate chains is *asbestos* (Figure 6-13). The fibrous character of asbestos is a direct consequence of the molecular structure of the $(Si_4O_{11}^{6-})_n$ polymeric chains.

The silicate minerals *mica* and *talc* contain two-dimensional, polymeric silicate sheets with the overall silicate composition $Si_2O_5^{2-}$. The structure of these sheets is illustrated in Figure 6-12(c), and Figure 6-14 shows how mica can easily be fractured into thin sheets. Talc has the composition $Mg_3(OH)_2(Si_2O_5)_2$, whereas micas have a variety of compositions, one example of which is *lepidolite*, $KLi_2Al(Si_2O_5)_2(OH)$. The ease with which mica can be separated into thin sheets and the slippery feel of talcum powder arise from the layered structure of the silicates in these minerals. Some commercially important compounds of silicon are given in Table 6-4.

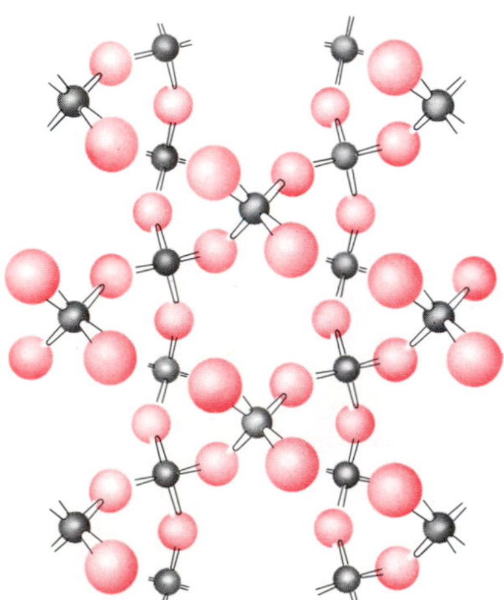

Figure 6-15 The crystalline structure of quartz. Note that each silicon atom is surrounded by four oxygen atoms. The silicon atoms are linked by the oxygen atoms.

6-7 MOST GLASSES ARE SILICATES

Quartz is a crystalline material with the composition SiO_2 and the crystalline structure shown in Figure 6-15. When crystalline quartz is melted and then cooled quickly to prevent the formation of crystals, there is formed a disordered three-dimensional array of polymeric chains, sheets, and other three-

Table 6-4 Some important compounds of silicon

Compound	*Uses*
fluosilicic acid, $H_2SiF_6(s)$	water fluoridation, sterilizing agent in the brewing industry; hardener in cement and ceramics
sodium silicate, $Na_4SiO_4(s)$	soaps and detergents; silica gels; adhesives; water treatment; sizing of textiles and paper; waterproofing cement; flame retardant, preservative
silicon carbide (carborundum), $SiC(s)$	abrasive for cutting and grinding metals
silicon dioxide (silica), $SiO_2(s)$	glass manufacture, abrasives, refractory material, cement
silicones, $\left(\text{—}\overset{\text{R}}{\underset{\text{R}'}{\text{Si}}}\text{—O—}\right)_n$	lubricants, adhesives, protective coatings, coolant, waterproofing agent, cosmetics, and many more

dimensional clusters. The resulting material is called quartz glass. All glass consists of a random array of these clusters.

Glass manufacturing is a 10-billion-dollar-per-year industry in the United States. The major component in glass is almost pure quartz sand. Among the other components of glass, soda (Na_2O) comes from soda ash (Na_2CO_3), lime (CaO) comes from limestone ($CaCO_3$), and aluminum oxide comes from feldspars, which have the general formula $M_2O{\cdot}Al_2O_3{\cdot}6SiO_2$, where M is K or Na. All the components of glass are fairly inexpensive chemicals.

A wide variety of glass properties can be produced by varying the glass composition. For example, partial replacement of CaO and Na_2O by B_2O_3 gives a glass that does not expand on heating or contract on cooling and is thus used in making glass utensils meant to be heated. Colored glass is made by adding a few percent of a colored transition metal oxide, such as CoO to make blue "cobalt" glass and Cr_2O_3 to make orange glass. Lead glass, which contains PbO, has attractive optical properties and is used to make decorative, cut-glass articles.

The addition of K_2O increases the hardness of glass and makes it easier to grind to precise shapes. Optical glass contains about 12 percent K_2O. Photochromic eyeglasses have a small amount of added silver chloride dispersed throughout and trapped in the glass. When sunlight strikes this type of glass, the tiny AgCl grains decompose into opaque clusters of silver atoms and chlorine atoms:

$$\underset{\text{clear}}{\text{AgCl}} \underset{\text{dark}}{\overset{\text{sunlight}}{\rightleftharpoons}} \underset{\text{opaque}}{\text{Ag} + \text{Cl}}$$

Steven Smale

Quartz often forms large, beautiful crystals.

▪ Silicon polishes, lubricants, and rubbers are silicon compounds with the following structure (X is variable):

$$
\begin{array}{ccccc}
 & CH_3 & & CH_3 & & CH_3 & \\
 & | & & | & & | & \\
CH_3- & Si & -O- & Si & -O- & Si & -CH_3 \\
 & | & & | & & | & \\
 & CH_3 & & CH_3 & & CH_3 & \\
\end{array}
$$

(the central $-Si(CH_3)_2-O-$ unit is enclosed in a dashed box with subscript X)

The chlorine atoms are trapped in the crystal lattice, and the silver and chlorine atoms recombine in the dark to form silver chloride, which causes the glass to become clear.

The etching of glass by hydrofluoric acid, $HF(aq)$, is a result of the reaction

$$SiO_2(s) + 6HF(aq) \rightarrow H_2SiF_6(s) + 2H_2O(l)$$

and this reaction is used to "frost" the inside surface of lightbulbs.

Porcelain has a much higher percentage of Al_2O_3 than glass and as a result is a heterogeneous substance. Porcelain is stronger than glass because of this hetereogeneity and is also more chemically resistant than glass. Earthenware is similar in composition to porcelain but is more porous because it is fired at a lower temperature.

6-8 GERMANIUM IS A SEMIMETAL; TIN AND LEAD ARE METALS

Germanium, a semimetal (Figure 6-7), is prepared in a manner similar to that for silicon. The main uses of germanium are in transistor technology and in infrared windows, prisms, and lenses. Germanium is transparent in the infrared.

Tin is found primarily in the mineral *cassiterite,* SnO_2, which occurs in rare but large deposits in Malaysia, China, the U.S.S.R., and the United States. Total U.S. natural reserves of tin are very

The minerals galena and cassiterite.

small. Tin is easily produced by heating SnO_2 with charcoal (carbon) and has been known since prehistoric times. It is used in plating (tin-plated food cans) and in various alloys, including solders, type metal, pewter, bronze, and gun metal. Tin objects are subject to a condition called *tin disease,* which is the conversion of the *white* allotrope of tin to the *gray* allotrope. This conversion occurs slowly below 13°C and results in the brittle gray allotrope of tin. A famous compound of tin is stannous (tin(II)) fluoride, SnF_2, a white crystalline powder that was the first fluoride additive used in toothpaste.

Lead is obtained primarily from the ore *galena,* PbS; commercial deposits occur in over 50 countries. The ore is first roasted in air:

$$2PbS(s) + 3O_2(g) \rightarrow 2PbO(s) + 2SO_2(g)$$

and then reduced by carbon in a blast furnace:

$$PbO(s) + CO(g) \rightarrow Pb(l) + CO_2(g)$$

Lead is resistant to corrosion and is used in a variety of alloys. Lead storage batteries constitute the major use of lead. The metal is also used in cable coverings, ammunition, and the synthesis of tetraethyl lead, $(CH_3CH_2)_4Pb$, which is used in leaded gasolines. Lead was once used in paints—$PbCrO_4$ is yellow and Pb_3O_4 is red—but lead salts constitute a serious health hazard, as they are cumulative poisons, and their use in paints has been discontinued. The Romans used lead vessels to store wine and other consumables and to conduct water in lead-lined aqueducts; thus lead poisoning may have had more to do with the collapse of the Roman Empire than any other factor. The use of lead-containing glazes on pottery for food use is now prohibited in the United States.

Table 6-5 Some important compounds of germanium, tin, and lead

Compound	*Uses*
germanium dioxide, $GeO_2(s)$	infrared-transmitting glass, transistors and diodes
tin(IV) oxide, $SnO_2(s)$	white enamels, ceramics and glass, polishing glass and marble, cosmetics
tin(II) chloride, $SnCl_2(s)$	reducing agent in dye manufacture, tin galvanizing, soldering flux
tin(II) fluoride, $SnF_2(s)$	toothpaste additive
tin(IV) chloride, $SnCl_4(s)$	perfume stabilization in soaps, ceramic coatings, manufacture of blueprint paper
lead(II) oxide, $PbO(s)$	glazing pottery and ceramics, lead glass
lead dioxide, $PbO_2(s)$	oxidizing agent, matches, lead-acid storage batteries, pyrotechnics
lead chromate, $PbCrO_4(s)$	yellow-to-red pigments
lead azide, $Pb(N_3)_2(s)$	detonating agent (primer)

The elements germanium, tin, and lead show a steady trend to metallic properties and to increasing stability of the +2 oxidation state. The +2 oxidation state is of little importance in the chemistry of germanium, but predominates for lead. The ion Pb^{4+} exists only in the solid state as, for example, in PbO_2. Lead has an extensive aqueous solution chemistry as the $Pb^{2+}(aq)$ ion. In addition, the salts of germanium, tin, and lead show increasing ionic character with increasing atomic number. For example, germanium reacts with the halogens upon moderate heating to form covalent tetrahalides of the form GeX_4. Tin forms both dihalides, SnX_2, and tetrahalides, SnX_4, which are also covalent. Lead reacts with the halogens to form dihalides, PbX_2, which have well-defined ionic character.

Tin and lead are mild reducing agents in aqueous media:

$$Sn^{2+}(aq) + 2e^- \rightleftharpoons Sn(s) \qquad E^0 = -0.14\ \text{V}$$
$$Pb^{2+}(aq) + 2e^- \rightleftharpoons Pb(s) \qquad E^0 = -0.13\ \text{V}$$

whereas lead(IV) oxide, $PbO_2(s)$, is a fairly strong oxidizing agent in acid solution:

$$PbO_2(s) + 4H^+(aq) + 2e^- \rightleftharpoons Pb^{2+}(aq) + 2H_2O(l) \qquad E^0 = 1.46\ \text{V}$$

This half-reaction occurs during discharge at the cathode (+ terminal) of the lead storage battery, except that the product is $PbSO_4(s)$ rather than $Pb^{2+}(aq)$.

The oxides become increasingly basic from germanium to lead. The oxide GeO_2 is slightly acidic, whereas the oxides SnO, SnO_2, and PbO and the hydroxides $Sn(OH)_2$ and $Pb(OH)_2$ are amphoteric. For example,

$$Pb(OH)_2(s) + 2H^+(aq) \rightleftharpoons Pb^{2+}(aq) + 2H_2O(l)$$
$$Pb(OH)_2(s) + 2OH^-(aq) \rightleftharpoons Pb(OH)_4^{2-}(aq)$$

As a consequence, water-insoluble lead(II) salts (except PbS) dissolve readily in strong base.

Tin and lead form a variety of organometallic compounds, such as $(CH_3CH_2)_2Sn$ and $(CH_3CH_2)_4Pb$ (tetraethyllead). Tetraethyllead is used as an antiknock compound in leaded gasolines because it promotes a smooth burning of the fuel, but its use is decreasingly rapidly because of EPA regulations. Table 6-5 lists some of the important compounds of germanium, tin, and lead.

TERMS YOU SHOULD KNOW

cassiterite
galena
allotropes
carborundum
zone refining
valence band
conduction band
n-type semiconductor
p-type semiconductor
orthosilicate ion
water glass
polymer
silicones
tin disease

QUESTIONS

6-1. Complete and balance the following equations.

(a) $Al_2O_3(s) + C(s) \xrightarrow[\text{furnace}]{\text{electric}}$

(b) $Al_4C_3(s) + D_2O(g) \xrightarrow{\text{heat}}$

(c) $CaC_2(s) + D_2O(l) \xrightarrow{\text{heat}}$

(d) $PbS(s) + O_2(g) \xrightarrow{\text{heat}}$

6-2. Complete and balance the following equations.

(a) $PbO(s) + CO(g) \xrightarrow{\text{heat}}$

(b) $Si(s) + NaOH(aq) \rightarrow$

(c) $SiCl_4(g) + H_2O(l) \rightarrow$

(d) $SiO_2(s) + 6HF(aq) \rightarrow$

6-3. Do the acidities of the Group 4 oxides increase or decrease with increasing atomic number?

6-4. Write chemical equations describing the amphoteric nature of $Sn(OH)_2$.

6-5. Given that graphite is converted to diamond under high pressures, use Le Châtelier's principle to deduce the relative densities of the two substances.

6-6. Explain on a molecular level why diamond is an extremely hard substance and graphite is slippery. Give applications of these properties.

6-7. Explain on a molecular level why diamond is a poor conductor of electricity and graphite is a good conductor.

6-8. Clean rainwater (as opposed to acid rain) is slightly acidic, with a pH of about 5.6. Explain the acidity of clean rainwater.

6-9. Calcium cyanamide, CaNCN, an important industrial chemical, is produced by the reaction

$$CaC_2(s) + N_2(g) \xrightarrow{1000^\circ C} CaNCN(s) + C(s)$$

Draw the Lewis formula of the cyanamide ion, and use VSEPR theory to predict its shape.

6-10. Cyanogen, $(CN)_2$, is a colorless, highly toxic gas that is used to synthesize many nitrogen-containing organic compounds. Draw the Lewis formula of cyanogen, and use VSEPR theory to predict its shape.

6-11. Based on analogy with the methods of preparation of other carbides, propose two ways to synthesize silicon carbide, SiC.

6-12. Give two examples (not mentioned in the text) of elements that can be added to make *n*-type and *p*-type silicon semiconductors, respectively.

6-13. Discuss and explain the difference in reactivities between the carbon tetrahalides and the silicon tetrahalides.

6-14. Determine the oxidation states of the Group 4 elements in the following compounds.

(a) SiC (b) CS_2 (c) NaCN (d) $Si_6O_{18}^{12-}$

6-15. Determine the oxidation state(s) of the Group 4 elements in the following compounds.

(a) PbO_2
(b) Pb_3O_4 ("red lead")
(c) Al_4C_3
(d) CaC_2

6-16. Draw structures for the following ions.

(a) $Si_2O_7^{6-}$ (b) $Si_3O_{10}^{8-}$ (c) $Si_6O_{18}^{12-}$

6-17. Write the chemical equation that describes the etching of glass by hydrofluoric acid.

6-18. Use VSEPR theory to predict the shapes of (a) $GeBr_4$, (b) $SnCl_2$, (c) GeF_6^{2-}, and (d) SiH_3^-.

6-19. Use hybrid orbitals to describe the bonding in the silanes, SiH_4 and Si_2H_6.

6-20. Write chemical equations describing how tin and lead are obtained from their ores.

6-21. Explain how photochromic glass works.

6-22. Many fine museum pieces and organ pipes made of tin have been ruined because their temperatures were allowed to drop below 13°C for appreciable periods of time. Explain.

6-23. How many kilograms of lead can be obtained from 100 kilograms of galena?

6-24. How many kilograms of tin can be obtained from 100 kilograms of cassiterite?

THE GROUP 5 ELEMENTS

The Group 5 elements. Back row from left to right: Nitrogen, phosphorus, and arsenic. Front row: Antimony and bismuth.

The Group 5 elements are nitrogen, phosphorus, arsenic, antimony, and bismuth. Nitrogen and phosphorus are nonmetals, arsenic and antimony are semimetals, and bismuth is a metal. Group 5 nicely illustrates the trend from acidic to basic oxides on descending a group. The oxides of nitrogen, phosphorus, and arsenic are acidic, those of antimony are amphoteric, and bismuth(III) oxide, Bi_2O_3, is basic. There is also an increase in stability of the lower oxidation state with increasing atomic number. Thus, Bi_2O_3 is the only stable oxide of bismuth, whereas the other members of the group also have oxides of the type M_2O_5. The Group 5 family gives us our first opportunity to

Table 7-1 Sources and uses of the Group 5 elements

Element	*Principal sources*	*Uses*
nitrogen	fractional distillation of liquid air	production of ammonia (Haber process), inert atmosphere for chemical processes, refrigerant (as liquid)
phosphorus	phosphate rock (impure $Ca_3(PO_4)_2$); apatites [$Ca_{10}F_2(PO_4)_6$, $Ca_{10}(OH)_2(PO_4)_6$]	manufacture of phosphoric acid, incendiaries, matches, pyrotechnics, smoke bombs
arsenic	*realgar* (As_4S_4), *orpiment* (As_2S_3), *mispickel* (FeAsS), flue dust of copper and lead smelters	doping agent in electronic devices, alloys with lead and copper, special solders
antimony	*stibnite* (Sb_2S_3)	infrared detectors, hardening alloy for lead, antifriction alloys, type metal, tracer bullets
bismuth	*bismuthinite* (Bi_2S_3), *bismite* (Bi_2O_3), by-product of lead, copper, and tin refining	low-melting alloys, pharmaceuticals, cosmetics, permanent magnets

note that the nonmetals of the fourth row, arsenic, selenium and bromine, favor an oxidation state less than their maximum. Thus, although AsF_3, $AsCl_3$, $AsBr_3$, and AsI_3 exist, only AsF_5 is known.

Nitrogen makes up almost 80 percent of the earth's atmosphere, from which it can be obtained by fractional distillation of liquefied air, a method that exploits the difference in the boiling points of nitrogen and oxygen, the principal components of air. The nitrogen can be separated from the oxygen because nitrogen boils at −196°C, whereas oxygen boils at −183°C. The lack of chemical reactivity of nitrogen accounts for its lack of abundance in mineral form in the earth's crust. The only significant nitrogen-containing minerals are KNO_3 (saltpeter) and $NaNO_3$ (Chile saltpeter). Phosphorus is the twelfth most abundant element in the earth's crust. All its important minerals are phosphates, which are collectively referred to as *phosphate rock.* Both nitrogen and phosphorus occur in all living things, making these two elements essential components of fertilizers. Arsenic, antimony, and bismuth are not particularly abundant. The principal sources and commercial uses of the Group 5 elements are given in Table 7-1.

Tables 7-2 and 7-3 present some atomic and physical properties of the Group 5 elements. Note that atomic size increases and that ionization energy and electronegativity decrease with increasing atomic number.

Table 7-2 Atomic properties of the Group 5 elements

Property	*N*	*P*	*As*	*Sb*	*Bi*
atomic number	7	15	33	51	83
atomic mass/amu	14.0067	30.97376	74.9216	121.75	208.9804
number of naturally occurring isotopes	2	1	1	2	1
ground-state electron configuration	$[He]2s^22p^3$	$[Ne]3s^23p^3$	$[Ar]3d^{10}4s^24p^3$	$[Kr]4d^{10}5s^25p^3$	$[Xe]4f^{14}5d^{10}6s^26p^3$
atomic radius/pm	65	100	115	145	160
ionization energy/$MJ \cdot mol^{-1}$					
first	1.40	1.06	0.947	0.834	0.701
second	2.86	1.90	1.70	1.60	1.61
third	4.58	2.91	2.74	2.44	2.47
Pauling electronegativity	3.0	2.1	2.0	1.9	1.8

7-1 NITROGEN, N_2, HAS A VERY STRONG TRIPLE BOND

Nitrogen is a colorless, odorless gas that exists as a diatomic molecule, N_2. The most significant property of elemental nitrogen is its lack of chemical reactivity. Nitrogen, as N_2, does not take part in many chemical reactions. The nitrogen molecule is generally unreactive because of the very high bond energy of the triple bond in N_2:

$$N{\equiv}N(g) \rightarrow 2N(g) \qquad \Delta H_{rxn} = 946 \text{ kJ}$$

Although nitrogen compounds are essential nutrients for animals and plants, only a few microorganisms are able to utilize elemental nitrogen directly by converting it to water-soluble compounds of nitrogen. The conversion of nitrogen from the free element to nitrogen compounds is one of the most important problems of modern chemistry and is called *nitrogen fixation.*

Table 7-3 Physical properties of the Group 5 elements

Property	N_2	*P (red)*	*P (white)*	*As*	*Sb*	*Bi*
melting point/°C	−210.0	—	44.1	814	631	271
boiling point/°C	−196.0	416 (sub)	280	613 (sub)	1440	1560
enthalpy of fusion/$kJ \cdot mol^{-1}$	0.72	—	2.51	21.3	19.8	10.9
enthalpy of vaporization/$kJ \cdot mol^{-1}$	5.58			—	193	151
density/$g \cdot cm^{-3}$	0.879	2.34	1.82	5.78	6.70	9.80 (rhomb)

Liquid nitrogen at its boiling point.

7-2 NITROGEN IS THE SECOND-RANKED INDUSTRIAL CHEMICAL

In terms of U.S. industrial production, nitrogen is the second leading chemical. Over 42 billion pounds of pure nitrogen is produced from air each year. Nitrogen is also found in potassium nitrate, KNO_3 (saltpeter), and in sodium nitrate, $NaNO_3$ (Chile saltpeter). Vast deposits of these two nitrates are found in the arid northern region of Chile, where there is insufficient rainfall to wash away these soluble compounds. The Chilean nitrate deposits are about 200 miles long, 20 miles wide, and many feet thick. At one time the economy of Chile was based primarily upon the sale of nitrates for use as fertilizers.

Large quantities of nitrogen are stored and shipped as the liquid in insulated metal cylinders. Smaller quantities are shipped as the gas in heavy-walled steel cylinders. The most convenient source of nitrogen gas in the laboratory is a steel cylinder charged with compressed N_2 gas. An alternative source is to heat an aqueous solution of ammonium nitrite, which thermally decomposes according to the equation

$$NH_4NO_2(aq) \rightarrow N_2(g) + 2H_2O(l)$$

Ammonium nitrite is a potentially explosive solid, and so the aqueous ammonium nitrite solution is made by adding ammonium chloride and sodium nitrite, both stable compounds, to water. Even so, the solution must be heated carefully to avoid an explosion.

Another laboratory preparation of nitrogen is the passage of ammonia gas over hot copper(II) oxide:

$$2NH_3(g) + 3CuO(s) \xrightarrow{400^\circ C} N_2(g) + 3H_2O(g) + 3Cu(s)$$

7-3 MOST NITROGEN IS CONVERTED TO AMMONIA BY THE HABER PROCESS

The inertness of nitrogen toward most other chemical substances makes reactions in which nitrogen combines with other elements economically important. Nitrogen fixation occurs both industrially and in nature. The most important industrial nitrogen-fixation reaction is the *Haber process,* in which nitrogen reacts directly with hydrogen at high pressure and high temperature to form ammonia (Figure 7-1):

$$N_2(g) + 3H_2(g) \xrightarrow[500^\circ C]{300\text{ atm}} 2NH_3(g)$$

Over 27 billion pounds of ammonia is produced annually in the United States by the Haber process. Rated in terms of pounds produced per year, ammonia is the sixth ranked industrial chemical in the United States.

Union 76

Figure 7-1 An ammonia plant. This plant produces 750 tons of ammonia per day from hydrogen gas and nitrogen gas. The nitrogen comes from air and the hydrogen is obtained from the reaction between methane and steam.

Ammonia is a colorless gas with a sharp, irritating odor. It is the effective agent in some forms of "smelling salts." Unlike nitrogen, ammonia is extremely soluble in water. Household ammonia is about a 2 M solution of NH_3 in water together with a detergent. Ammonia was the first complex molecule to be identified in interstellar space. Ammonia occurs in galactic dust clouds in the Milky Way and, in the solid form, constitutes the rings of Saturn.

Nitrogen reacts directly with lithium metal at room temperature to form lithium nitride:

$$6Li(s) + N_2(g) \rightarrow 2Li_3N(s)$$

The reddish-black lithium nitride reacts directly with water to form ammonia:

$$Li_3N(s) + 3H_2O(l) \rightarrow 3LiOH(aq) + NH_3(g)$$

This reaction can be used to prepare deuterated ammonia, ND_3:

$$Li_3N(s) + 3D_2O(l) \rightarrow 3LiOD(s) + ND_3(g)$$

Preparation of ammonia by these reactions is not competitive economically with the Haber process because of the high cost of producing lithium metal.

7-4 FIXED NITROGEN IS A KEY INGREDIENT IN FERTILIZERS

Ammonia is readily soluble in water, binds to many components of soil, and is easily converted to usable plant food. Concentrated aqueous solutions of ammonia or pure liquid ammonia can be sprayed directly into the soil (Figure 7-2). Ammonia is inexpensive and high in nitrogen. The increased growth of plants when fertilized by ammonia is spectacular. Liquid ammonia is toxic and injurious to living tissue, however, and must be handled carefully.

For some purposes it is more convenient to use a solid fertilizer instead of ammonia solutions. For example, ammonia combines directly with sulfuric acid to produce ammonium sulfate:

$$2NH_3(aq) + H_2SO_4(aq) \rightarrow (NH_4)_2SO_4(aq)$$

Ammonium sulfate is the most important solid fertilizer in the world. Its annual U.S. production is 4 billion pounds.

The primary fertilizer nutrients are nitrogen, phosphorus, and potassium, and fertilizers are rated by how much of each they contain. For example, a 5-10-5 fertilizer has 5 percent by mass total available nitrogen, 10 percent by mass phosphorus (equivalent to the form P_2O_5), 5 percent by mass potassium (equivalent to the form K_2O), and 80 percent inert ingredients. The production of fertilizers is one of the largest and most important industries in the world.

Grant Heilman/Grant Heilman

Figure 7-2 This photo demonstrates the method of spraying ammonia into the soil. Liquid ammonia, called anhydrous ammonia, is used extensively as a fertilizer because it is cheap, high in nitrogen, and easy to apply.

7-5 NITRIC ACID IS PRODUCED BY THE OSTWALD PROCESS

About half of all the ammonia produced is converted to nitric acid by the *Ostwald process*. The first step in this process is the conversion of ammonia to nitrogen oxide:

$$4NH_3(g) + 5O_2(g) \xrightarrow[825°C]{Pt} 4NO(g) + 6H_2O(g) \qquad (1)$$

The second step in the Ostwald process involves the oxidation of NO to nitrogen dioxide by reaction with oxygen:

$$2NO(g) + O_2(g) \rightarrow 2NO_2(g) \qquad (2)$$

In the final step, the NO_2 is dissolved in water to yield nitric acid:

$$3NO_2(g) + H_2O(l) \rightarrow 2HNO_3(aq) + NO(g) \qquad (3)$$

The $NO(g)$ evolved is recycled back to step (2).

Laboratory grade nitric acid is approximately 70 percent HNO_3 by weight with a density of $1.42\ g \cdot mL^{-1}$ and a concentration of 16 M (Figure 7-3). The U.S. annual production of nitric acid is over 14 billion pounds, which makes it the eleventh-ranked industrial chemical. Nitric acid is the least expensive potent oxidizing agent and is used in a large number of important chemical processes, including the production of explosives such as trinitrotoluene (TNT), nitroglycerine, and nitrocellulose (gun cotton). It is also used in etching and photoengraving processes to produce grooves in metal surfaces.

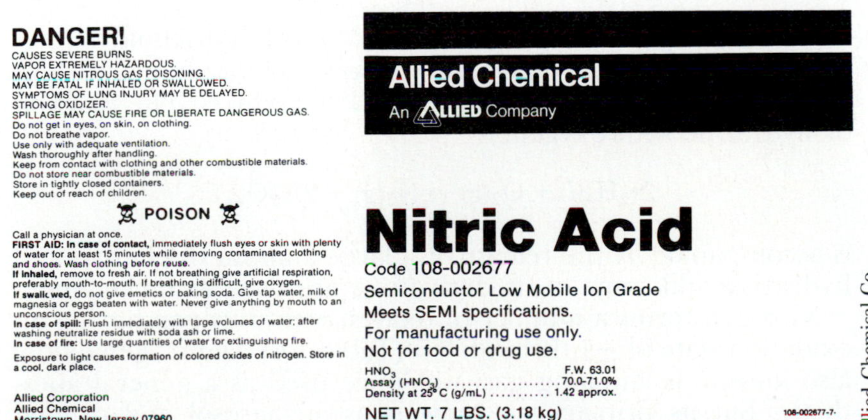

Figure 7-3 Label from a bottle of concentrated nitric acid. Notice that the label contains information on the hazardous properties of the substance.

7-6 CERTAIN BACTERIA CAN FIX NITROGEN

USDA

Figure 7-4 Nitrogen-fixing nodules on the roots of a leguminous plant. The nodules contain *Rhizobium,* a soil bacterium that converts atmospheric elemental nitrogen to water-soluble nitrogen compounds.

Nitrogen fixation by microorganisms is an important source of plant nutrients. The most common of these nitrogen-fixing bacteria is the *Rhizobium* bacterium, which invades the roots of leguminous plants, such as alfalfa, clover, beans, and peas. The *Rhizobium* forms nodules on the roots of these legumes and has a symbiotic (mutually beneficial) relationship with the plant (Figure 7-4). The plant produces carbohydrates through photosynthesis, and the *Rhizobium* uses the carbohydrate as fuel for fixing the nitrogen, which is incorporated into plant protein. Alfalfa is the most potent nitrogen-fixer, followed by clover, soybeans, other beans, peas, and peanuts. In modern agriculture, crops are rotated, meaning that a nonleguminous crop and a leguminous crop are alternated on one piece of land. The leguminous crop is either harvested, leaving behind nitrogen-rich roots, or plowed into the soil, adding both nitrogen and organic matter. A plowed-back crop of alfalfa may add as much as 400 lb of fixed nitrogen to the soil per acre.

7-7 NITROGEN FORMS SEVERAL IMPORTANT COMPOUNDS WITH HYDROGEN AND OXYGEN

The most important nitrogen-hydrogen compounds are ammonia, NH_3, hydrazine, N_2H_4, and hydrazoic acid, HN_3. Ammonia is a weak base. The pK_b of the reaction

$$NH_3(aq) + H_2O(l) \rightleftharpoons NH_4^+(aq) + OH^-(aq)$$

is 4.76 at 25°C. Hydrazine is a colorless, fuming, reactive liquid. It is produced by the *Raschig synthesis,* in which ammonia is reacted with hypochlorite ion (household bleach is sodium hypochlorite in water) in basic solution:

$$2NH_3(aq) + ClO^-(aq) \xrightarrow{OH^-(aq)} N_2H_4(aq) + H_2O(l) + Cl^-(aq)$$

Household bleach should *never* be mixed with household ammonia because extremely toxic and explosive chloramines, such as H_2NCl and $HNCl_2$, are produced as by-products. The reaction of hydrazine with oxygen,

$$N_2H_4(l) + O_2(g) \rightarrow N_2(g) + 2H_2O(g)$$

is accompanied by the release of a large amount of energy, and hydrazine and some of its derivatives are used as rocket fuels.

Nitrogen forms a number of oxides, with nitrogen having an oxidation state of +1 through +5 (Table 7-4). Dinitrogen oxide, also known as laughing gas, was once used as a general anesthetic, but its primary use now is as an aerosol and canned whipped cream propellant. Dinitrogen oxide can be produced by a cautious thermal decomposition of NH_4NO_3:

$$NH_4NO_3(s) \rightarrow N_2O(g) + 2H_2O(l)$$

Table 7-4 The principal oxides of nitrogen

Formula	*Systematic name*	*Description*
N_2O	dinitrogen oxide (nitrous oxide)	colorless, rather unreactive gas
NO	nitrogen oxide (nitric oxide)	colorless, paramagnetic, reactive gas
N_2O_3	dinitrogen trioxide	dark-blue solid (m.p. −101°C); dissociates in gas phase to NO and NO_2
NO_2	nitrogen dioxide	brown, paramagnetic, reactive gas; dimerizes reversibly to N_2O_4
N_2O_4	dinitrogen tetroxide	colorless gas (b.p. 21°C) dissociates reversibly to NO_2
N_2O_5	dinitrogen pentoxide	colorless, ionic solid; unstable as a gas

Nitrogen oxide is produced in the oxidation of copper by dilute nitric acid:

$$3Cu(s) + 8HNO_3(aq) \rightarrow 3Cu(NO_3)_2(aq) + 2NO(g) + 4H_2O(l)$$

Although NO(*g*) is colorless, this reaction appears to produce a brown gas if it is run in a vessel that is open to the atmosphere. The brown gas results from the rapid production of nitrogen dioxide by the reaction

$$\underset{\text{colorless}}{2NO(g)} + O_2(g) \rightarrow \underset{\text{brown}}{2NO_2(g)}$$

In the gas phase, nitrogen dioxide dimerizes to form dinitrogen tetroxide:

$$2NO_2(g) \rightleftharpoons N_2O_4(g) \qquad \Delta H^\circ_{rxn} = -57.2\ \text{kJ} \cdot \text{mol}^{-1}$$

Because this reaction is exothermic, an increase in temperature results in the formation of more $NO_2(g)$, and hence a more reddish-brown mixture.

Dinitrogen trioxide can be prepared by the reaction

$$NO(g) + NO_2(g) \rightleftharpoons N_2O_3(g) \qquad \Delta H^\circ_{rxn} = -39.7\ \text{kJ} \cdot \text{mol}^{-1}$$

Because the reaction is exothermic, production of $N_2O_3(g)$ is favored at lower temperatures. Dinitrogen trioxide is formally the acid anhydride of nitrous acid, HNO_2, which can be prepared by the reaction of an equimolar mixture of nitrogen oxide and nitrogen dioxide in a basic solution (for example, NaOH):

$$NO(g) + NO_2(g) + 2NaOH(aq) \rightarrow 2NaNO_2(aq) + H_2O(l)$$

Addition of acid to the resulting solution yields nitrous acid:

$$NO_2^-(aq) + H^+(aq) \rightarrow HNO_2(aq)$$

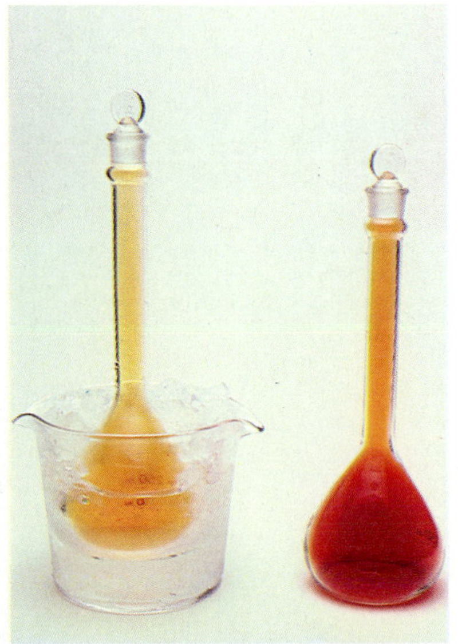

An increase in temperature from 0°C (ice water) to 25°C converts some of the N_2O_4 to NO_2 and results in a darker color for the reaction mixture.

Detonation ("blasting") caps.

Nitrous acid is a weak acid, with $pK_a = 5.22$ at 25°C.

Salts of nitrous acid are called nitrites. Sodium nitrite, $NaNO_2$, is used as a meat preservative. The nitrite ion combines with the hemoglobin in meat to produce a deep red color. The main problem with the extensive use of nitrites in foods is that the nitrite ion reacts with amines in the body's gastric juices to produce compounds called nitrosamines, such as $(CH_3)_2NNO$, dimethylnitrosamine, which are carcinogenic.

Dinitrogen pentoxide is the anhydride of nitric acid:

$$N_2O_5(s) + H_2O(l) \rightarrow 2HNO_3(aq)$$

Dinitrogen pentoxide is a rather unstable ionic solid and a powerful oxidizing agent.

The reaction of nitrous acid with hydrazine in acidic solution yields hydrazoic acid:

$$N_2H_4(aq) + HNO_2(aq) \rightarrow HN_3(aq) + 2H_2O(l)$$

Hydrazoic acid is a colorless, toxic liquid and a dangerous explosive. In aqueous solution, HN_3 is a weak acid, with $pK_a = 4.72$ at 25°C. Its lead and mercury salts, $Pb(N_3)_2$ and $Hg(N_3)_2$, which are called *azides*, are used in detonation caps; both compounds are dangerously explosive. Sodium azide, NaN_3, is used as the gas source in automobile air safety bags, which inflate rapidly on impact. Some other commercially important nitrogen-containing compounds are given in Table 7-5.

Table 7-5 Some important compounds of nitrogen

Compound	*Uses*
ammonia, NH_3	fertilizers; manufacture of nitric acid, explosives; synthetic fibers; refrigerant; manufacture of dyes and plastics
nitric acid, HNO_3	manufacture of fertilizers, explosives, lacquers, synthetic fabrics, drugs, and dyes; oxidizing agent; metallurgy; ore flotation
ammonium nitrate, NH_4NO_3	fertilizer, explosives, pyrotechnics, herbicides and insecticides, solid rocket propellant
calcium cyanamide, $CaCN_2$	fertilizer, herbicide, defoliant, hardening of iron and steel
sodium cyanide, NaCN	extraction of gold and silver from their ores; electroplating; case-hardening of metals; insecticide; fumigant; manufacture of dyes and pigments; ore flotation

7-8 OXIDES OF NITROGEN ARE THE PRIMARY INGREDIENTS OF PHOTOCHEMICAL SMOG

Under ordinary conditions, nitrogen and oxygen do not react with each other. When combined at high pressure and temperature, however, as in the cylinders of an automobile engine, they react to form nitrogen oxide, NO, which then reacts with O_2 to produce nitrogen dioxide. Ordinarily this reaction occurs too slowly at the low concentrations of NO in the atmosphere to account for any significant concentration of $NO_2(g)$, but, for reasons that are not yet understood, the reaction occurs rapidly in sunny, urban atmospheres. Nitrogen dioxide is a red-brown noxious gas that is responsible for the yellow-brown color of smog, first made famous in Los Angeles but now all too common in many urban areas. A concentration of 500 ppm of NO_2 in air is usually fatal; there is some disagreement concerning tolerable levels of NO_2, but they are not higher than 3 to 5 ppm. Levels of NO_2 reach 0.9 ppm in Los Angeles on particularly bad days.

The problem of NO_2 is not so much its primary toxicity but the fact that it is dissociated by radiation to produce atomic oxygen:

$$NO_2(g) \xrightarrow{\text{392-nm light}} NO(g) + O(g)$$

Because the dissociation of the NO_2 is caused by radiation (light), it is called *photodissociation.* The atomic oxygen then reacts with molecular oxygen to produce ozone. These two reactions account for the fact that ozone levels are higher on sunny days than on cloudy days. Ozone in the atmosphere makes up about 90 percent of the general category of pollutants called oxidants, which are now measured continually in many cities. Los Angeles has air pollution alerts when the level of oxidants exceeds 0.35 ppm.

The atomic oxygen produced by the photodissociation of NO_2 also attacks the hydrocarbons introduced into the atmosphere by the incomplete combustion of gasoline and diesel fuel. The reaction of atomic oxygen with hydrocarbons initiates a complicated sequence of chemical reactions. The end products of these reactions are a number of substances that attack living tissue and lead to great discomfort, if not serious disorders. These substances make up what is called *photochemical smog,* which causes eyes to tear and smart, something that people who live in smoggy cities experience often.

The control of photochemical smog requires controlling the emission of its two principal ingredients, NO and hydrocarbons, from automobile exhausts. The Congressional Clean Air Act of 1967, with its amendments in 1970 and 1977, imposed limitations on exhaust emissions. Although there are indications that smog has lessened in some cities, in many others smog and other types of pollution problems are still increasing.

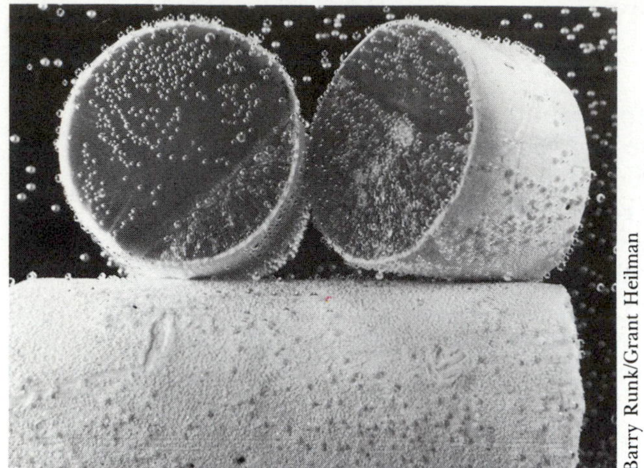

Barry Runk/Grant Heilman

Figure 7-5a White phosphorus is stored in water, as it ignites in air at about 25°C.

Figure 7-5b Red phosphorus is less reactive than white phosphorus, undergoing the same chemical reactions but at higher temperatures.

7-9 THERE ARE TWO PRINCIPAL ALLOTROPES OF SOLID PHOSPHORUS

There are several allotropic forms of solid phosphorus, the most important of which are *white phosphorus* and *red phosphorus.* White phosphorus is a white, transparent, waxy crystalline solid (Figure 7-5) that often appears pale yellow because of impurities. It is insoluble in water and alcohol but soluble in carbon disulfide. A characteristic property of white phosphorus is its high chemical reactivity. It ignites spontaneously in air at about 25°C. White phosphorus should never be allowed to come into contact with the skin because body temperature (37°C) is sufficient to ignite it spontaneously. Phosphorus burns are extremely painful and slow to heal. In addition, white phosphorus is very poisonous. White phosphorus should always be kept under water and handled with forceps.

When white phosphorus is heated above 250°C in the absence of air, a form called red phosphorus is produced. Red phosphorus is a red to violet powder that is less reactive than white phosphorus. The chemical reactions that the red form undergoes are the same as those of the white form, but they generally occur only at higher temperatures. For example, red phosphorus must be heated to 260°C before it burns in air. The toxicity of red phosphorus is much lower than that of white phosphorus.

White phosphorus consists of tetrahedral P_4 molecules (Figure 7-6), whereas red phosphorus consists of large, random aggregates of phosphorus atoms. The structure of red phosphorus is called *amorphous,* which means that it has no definite shape.

Figure 7-6 White phosphorus consists of tetrahedral P_4 molecules.

Most of the phosphorus that is produced is used to make phosphoric acid or other phosphorus compounds. Elemental phosphorus, however, is used in the manufacture of pyrotechnics, matches, rat poisons, incendiary shells, smoke bombs, and tracer bullets.

Phosphorus is not found as the free element in nature. The principal sources are the minerals calcium phosphate, hydroxyapatite, (Figure 7-7), fluorapatite, and chlorapatite. These ores collectively are called phosphate rock. Vast phosphate rock deposits occur in the U.S.S.R., in Morocco, and in Florida, Tennessee, and Idaho. An electric furnace is used to obtain phosphorus from phosphate rock. The furnace is charged with powdered phosphate rock, sand (SiO_2), and carbon in the form of coke. The source of heat is an electric current that produces temperatures of over 1000°C. A simplified version of the overall reaction that takes place is

▪ $Ca_3(PO_4)_2$	calcium phosphate
$Ca_{10}(OH)_2(PO_4)_6$	hydroxyapatite
$Ca_{10}F_2(PO_4)_6$	fluorapatite
$Ca_{10}Cl_2(PO_4)_6$	chlorapatite

$$\underset{\text{phosphate rock}}{2Ca_3(PO_4)_2(s)} + \underset{\text{sand}}{6SiO_2(s)} + \underset{\text{coke}}{10C(s)} \rightarrow 6CaSiO_3(l) + 10CO(g) + P_4(g)$$

The liquid calcium silicate, $CaSiO_3(l)$, called slag, is tapped off from the bottom of the furnace, and the phosphorus vapor produced solidifies to the white solid when the mixture of $CO(g)$ and $P_4(g)$ is passed through water (carbon monoxide is only very slightly soluble in water).

Although some phosphate rock is used to make elemental phosphorus, most phosphate rock is used in the production of fertilizers. Phosphorus is a required nutrient of all plants, and phosphorus compounds have long been used as fertilizer. In

Bill Tronca/Tom Stack and Assoc.

Figure 7-7 Hydroxyapatite is a common mineral that is found in many areas. It is used in the manufacture of phosphoric acid and phosphate fertilizers.

spite of its great abundance, phosphate rock cannot be used as a fertilizer because, as the name implies, it is insoluble in water. Consequently, plants are not able to assimilate the phosphorus from phosphate rock. To produce a water-soluble source of phosphorus, phosphate rock is reacted with sulfuric acid to produce a water-soluble product called *superphosphate,* $Ca(H_2PO_4)_2$, one of the world's most important fertilizers.

7-10 THE OXIDES OF PHOSPHORUS ARE ACID ANHYDRIDES

The main difference in the chemistries of nitrogen and phosphorus is similar to that between oxygen and sulfur, namely, the availability of $3d$ orbitals on phosphorus. The $3d$ orbitals make possible the expansion of the valence shell beyond the octet and thus the occurrence of more than four bonds to phosphorus.

White phosphorus reacts directly with oxygen to produce the oxides P_4O_6 and P_4O_{10}. With excess phosphorus present, P_4O_6 is formed:

$$\underset{\text{excess}}{P_4(s)} + 3O_2(g) \rightarrow P_4O_6(s)$$

with excess oxygen present, P_4O_{10} is formed:

$$P_4(s) + \underset{\text{excess}}{5O_2(g)} \rightarrow P_4O_{10}(s)$$

The formulas for P_4O_6 and P_4O_{10} are often written P_2O_3 and P_2O_5, respectively. These obsolete (that is, now known to be incorrect) molecular formulas are the basis for the common names phosphorus *tri*oxide and phosphorus *pent*oxide.

It is interesting to compare the structures of P_4O_6 and P_4O_{10} (Figure 7-8). The structure of P_4O_6 is obtained from that of P_4 by inserting an oxygen atom between each pair of adjacent phosphorus atoms; there are six edges on a tetrahedron, and thus a total of six oxygen atoms are required. The structure of P_4O_{10} is obtained from that of P_4O_6 by attaching an additional oxygen atom to each of the four phosphorus atoms.

▪ P_4O_6 produces a violent explosion when added to hot water.

The phosphorus oxides P_4O_6 and P_4O_{10} react with water to form the phosphorus oxyacids: phosphorous acid, H_3PO_3, and phosphoric acid, H_3PO_4:

$$P_4O_6(s) + 6H_2O(l) \rightarrow 4H_3PO_3(aq)$$
$$P_4O_{10}(s) + 6H_2O(l) \rightarrow 4H_3PO_4(aq)$$

Phosphorus pentoxide is a powerful dehydrating agent capable of removing water from concentrated sulfuric acid, which is itself a strong dehydrating agent.

$$P_4O_{10}(s) + 6H_2SO_4(l) \rightarrow 4H_3PO_4(l) + 6SO_3(g)$$

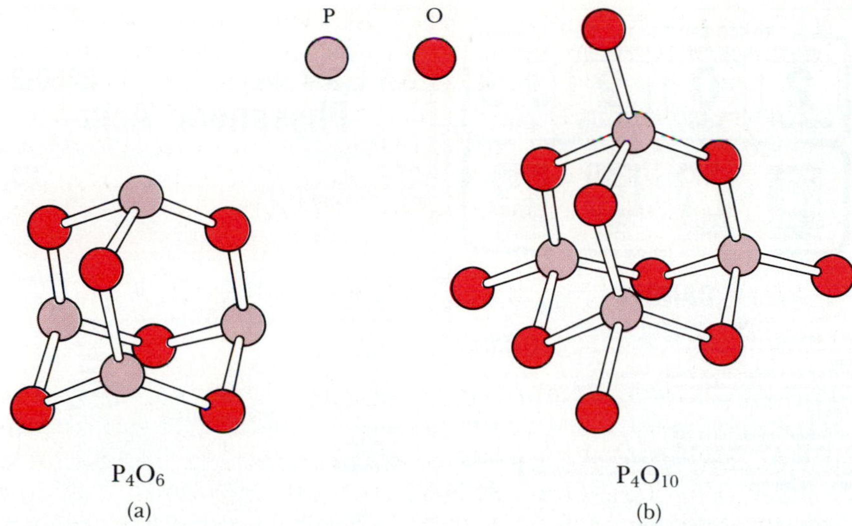

Figure 7-8 Structure of P_4O_6 and P_4O_{10}. (a) The P_4O_6 molecule can be viewed as arising from the tetrahedral P_4 molecule when an oxygen atom is inserted between each pair of adjacent phosphorus atoms. (b) The P_4O_{10} molecule can be viewed as arising from P_4O_6 when an oxygen atom is attached to each of the four phosphorus atoms. Note that there are no phosphorus-phosphorus bonds in either P_4O_6 or P_4O_{10}.

Thus phosphorus pentoxide is used as a drying agent in desiccators and dry boxes to remove water vapor.

Hypophosphorous acid, H_3PO_2, is prepared by reacting $P_4(g)$ with a warm aqueous solution of NaOH:

$$P_4(g) + 3OH^-(aq) + 3H_2O(l) \rightarrow 3H_2PO_2^-(aq) + PH_3(g)$$

followed by acidification:

$$H_2PO_2^-(aq) + H^+(aq) \rightarrow H_3PO_2(aq)$$

The structures of the phosphate anion, PO_4^{3-}, the phosphite ion, HPO_3^{2-}, and the hypophosphite ion, $H_2PO_2^-$, are

$$\left[\begin{matrix} O & & O \\ & P & \\ O & & O \end{matrix}\right]^{3-} \quad \left[\begin{matrix} H & & O \\ & P & \\ O & & O \end{matrix}\right]^{2-} \quad \left[\begin{matrix} H & & H \\ & P & \\ O & & O \end{matrix}\right]^{-}$$

Using VSEPR theory, we predict that these ions are tetrahedral, which is correct.

The hydrogen atoms attached to phosphorus are not dissociable in aqueous solutions. Thus, phosphoric acid, H_3PO_4, is triprotic, phosphorus acid, $H_2(HPO_3)$, is diprotic, and hypophosphorous acid, $H(H_2PO_2)$, is monoprotic. The pK_a values at 25°C are given in Table 7-6.

Table 7-6 The pK_a values in water at 25°C of phosphorus oxyacids

Acid	pK_{a1}	pK_{a2}	pK_{a3}
H_3PO_4	2.2	7.1	12.4
$H_2(HPO_3)$	1.8	~7	none
$H(H_2PO_2)$	1.2	none	none

Phosphoric acid (Figure 7-9) is the ninth ranked industrial chemical, almost 22 billion pounds being produced annually in

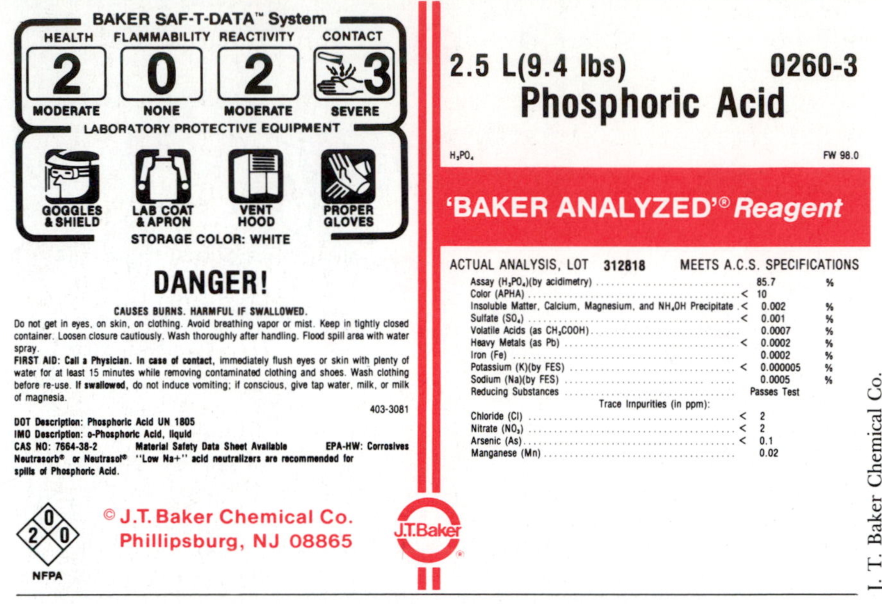

J. T. Baker Chemical Co.

Figure 7-9 Phosphoric acid label. Phosphoric acid is sold as a 15 M aqueous solution.

the United States. It is sold in various concentrations. The 85 percent by mass (85 g of H_3PO_4 to 15 g of H_2O) solution is a colorless, syrupy liquid. Laboratory phosphoric acid is sold as an 85% solution, which is equivalent to 15 M.

Phosphoric acid is used extensively in the production of soft drinks, and various of its salts are used in the food industry. For example, the monosodium salt, NaH_2PO_4, is used in a variety of foods to control acidity, and calcium dihydrogen phosphate, $Ca(H_2PO_4)_2$, is the acidic ingredient in baking powder. The evolution of carbon dioxide that takes place when baking powder is heated can be represented as

$$\underbrace{Ca(H_2PO_4)_2(s) + 2NaHCO_3(s)}_{\text{in baking powder}} \xrightarrow{300^\circ C} 2CO_2(g) + 2H_2O(g) + CaHPO_4(s) + Na_2HPO_4(s)$$

▪ Unleavened bread does not contain baking powder and thus does not rise when baked.

The slowly evolving $CO_2(g)$ gets trapped in small gas pockets and thereby causes the cake or bread to rise.

When phosphoric acid is heated gently, pyrophosphoric acid (pyro means heat) is obtained as a result of the elimination of a water molecule from a pair of phosphoric acid molecules:

$$\text{HO—}\overset{\text{O}}{\underset{\text{OH}}{\overset{\|}{\underset{|}{\text{P}}}}}\text{—}\boxed{\text{OH + H}}\text{O—}\overset{\text{O}}{\underset{\text{H}}{\overset{\|}{\underset{|}{\text{P}}}}}\text{—OH} \rightarrow \text{HO—}\overset{\text{O}}{\underset{\text{OH}}{\overset{\|}{\underset{|}{\text{P}}}}}\text{—O—}\overset{\text{O}}{\underset{\text{OH}}{\overset{\|}{\underset{|}{\text{P}}}}}\text{—OH} + H_2O$$

elimination of water

pyrophosphoric acid, $H_4P_2O_7$

Pyrophosphoric acid is a viscous, syrupy liquid that tends to solidify on long standing. In aqueous solution, it slowly reverts to phosphoric acid.

Longer chains of phosphate groups can be formed. The compound sodium tripolyphosphate, $Na_4P_3O_{10}$, used to be the phosphate ingredient of detergents. Its role was to break up and suspend dirt and stains by forming water-soluble complexes with metal ions. (The formation of complexes is discussed in Chapter 23 of the text.) In the 1960s almost all detergents contained phosphates, sometimes as much as 50 percent by mass. It was discovered, however, that the phosphates led to a serious water pollution problem. The enormous quantity of phosphates discharged into rivers and lakes served as a nutrient for the rampant growth of algae and other organisms. When these organisms died, much of the oxygen dissolved in the water was consumed in the decay process, thus depleting the water's oxygen supply and destroying the ecological balance. This process is called *eutrophication.* As a result of legislation in the 1970s, phosphates have been eliminated from detergents or their levels have been reduced markedly.

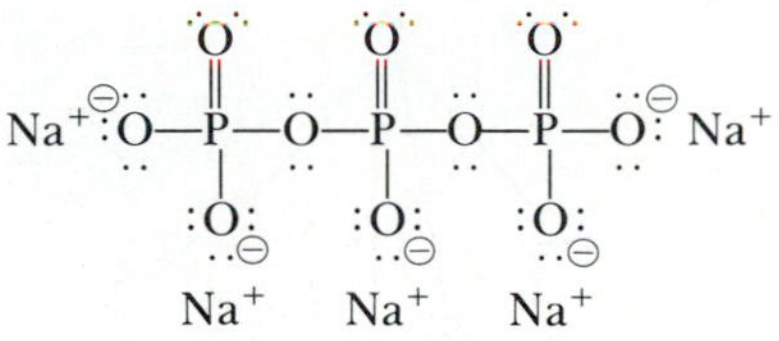

$Na^+\,{}^-O$—P(=O)(O$^-$ Na^+)—O—P(=O)(O$^-$ Na^+)—O—P(=O)(O$^-$ Na^+)—$O^-\,Na^+$

Sodium tripolyphosphate

7-11 PHOSPHORUS FORMS A NUMBER OF BINARY COMPOUNDS

Phosphorus reacts directly with reactive metals, such as sodium and calcium, to form phosphides. A typical reaction is

▪ Note the similarity of phosphides to nitrides.

$$12Na(s) + P_4(s) \rightarrow 4Na_3P(s)$$

Most metal phosphides react vigorously with water to produce phosphine, $PH_3(g)$:

$$Ca_3P_2(s) + 6H_2O(l) \rightarrow 2PH_3(g) + 3Ca(OH)_2(aq)$$

Phosphine has a trigonal pyramidal structure with an H—P—H bond angle of 93.7°. It is a colorless, extremely toxic gas with an offensive odor like that of rotten fish. Unlike ammonia, phosphine does not act as a base toward water, and few phosphonium (PH_4^+) salts are stable. Phosphine can also be prepared by the reaction of white phosphorus with a strong base.

▪ Phosphine reacts violently with oxygen and the halogens.

Phosphorus reacts directly with the halogens to form halides of the form PX_3 and PX_5. If an excess of phosphorus is used, then the trihalide is formed. For example,

$$\underset{\text{excess}}{P_4(s)} + 6Cl_2(g) \rightarrow 4PCl_3(l)$$

If an excess of halide is used, then the pentahalide is formed:

$$P_4(s) + \underset{\text{excess}}{10Cl_2(g)} \rightarrow 4PCl_5(s)$$

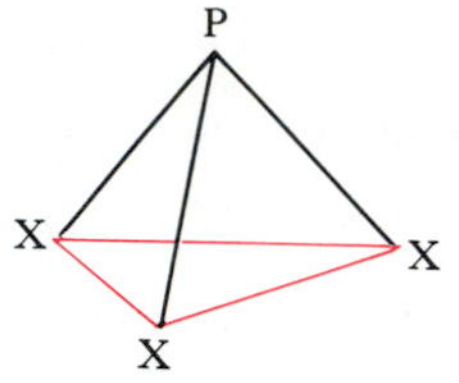

Figure 7-10 The phosphorus trihalides, PX_3, have a trigonal pyramidal structure in the gas phase.

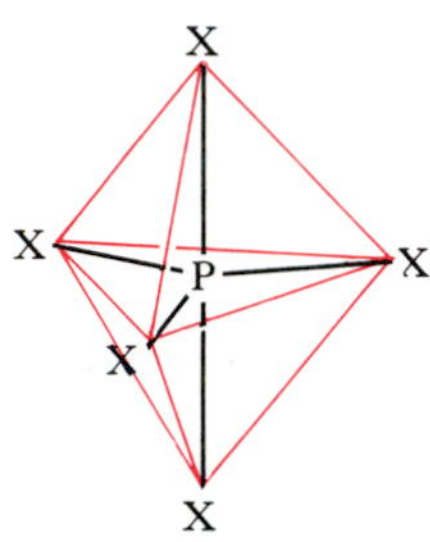

Figure 7-11 The phosphorus pentahalides, PX_5, have a trigonal bipyramidal structure in the gas phase.

Recall from Chapter 11 of the text that phosphorus trihalide molecules in the gas phase have a trigonal pyramidal structure (Figure 7-10) and that phosphorus pentahalide molecules in the gas phase have a trigonal bipyramidal structure (Figure 7-11). Table 7-7 lists the physical states of the various phosphorus halides. Phosphorus halides react vigorously with water. For example,

$$PCl_3(l) + 3H_2O(l) \rightarrow H_3PO_3(aq) + 3HCl(aq)$$
$$PCl_5(s) + 4H_2O(l) \rightarrow H_3PO_4(aq) + 5HCl(aq)$$

These hydrolysis reactions of PCl_3 and PCl_5 are fairly typical for molecular halides in which the central atom can bond to more atoms. The products are the hydrohalic acid and an oxyacid of the central atom. We encountered reactions of this type earlier (Sections 5-2 and 6-2):

$$BCl_3(g) + 3H_2O(l) \rightarrow B(OH)_3(s) + 3HCl(aq)$$
$$SiF_4(g) + 2H_2O(l) \rightarrow SiO_2(s) + 4HF(aq)$$

Note that the oxidation state of the central atom does not change in these reactions. The intermediates in these cases are believed to be species such as

$$\begin{array}{c} Cl \\ | \\ Cl-\overset{\ominus}{\ddot{P}}-Cl \\ | \\ {}^{\oplus}OH_2 \end{array} \quad \text{and} \quad \begin{array}{c} Cl \\ | \\ HO-\overset{\ominus}{\ddot{P}}-OH \\ | \\ {}^{\oplus}OH_2 \end{array}$$

If the central atom has its maximum number of bonds, as in CF_4 or SF_6, such intermediate species cannot form and the halides do not undergo hydrolysis.

Table 7-7 Physical states of the phosphorus halides at room temperature

Halide	*Physical state*	*Molecular description*
PF_3	colorless gas	trigonal pyramidal PF_3 molecules
PCl_3	clear, colorless, fuming liquid	trigonal pyramidal PCl_3 molecules
PBr_3	colorless, fuming liquid	trigonal pyramidal PBr_3 molecules
PI_3	red, crystalline, unstable solid	trigonal pyramidal PI_3 molecules
PF_5	colorless gas	trigonal bipyramidal PF_5 molecules
PCl_5	pale-yellow, fuming crystals	$[PCl_4^+][PCl_6^-]$ ion-pairs; PCl_4^+ tetrahedral and PCl_6^- octahedral
PBr_5	yellow, fuming, hygroscopic crystals	$[PBr_4^+]Br^-$ ion-pairs; PBr_4^+ tetrahedral
PI_5	not known	presumably iodine atoms are too large to arrange more than three around a phosphorus atom

When phosphorus is heated with sulfur, the yellow crystalline compound tetraphosphorus trisulfide, P_4S_3, is formed. Matches that can be ignited by striking on any rough surface contain a tip composed of the yellow P_4S_3 on top of a red portion that contains lead dioxide, PbO_2, together with antimony sulfide, Sb_2S_3. Friction causes the P_4S_3 to ignite in air, and the heat produced then initiates a reaction between antimony sulfide and lead dioxide that produces a flame.

Safety matches consist of a mixture of potassium chlorate and antimony sulfide. The match is ignited by striking on a special rough surface composed of a mixture of red phosphorus, glue, and abrasive. The red phosphorus is ignited by friction and in turn ignites the reaction mixture in the matchhead.

7-12 MANY PHOSPHORUS COMPOUNDS ARE IMPORTANT BIOLOGICALLY

Many organic phosphates are potent insecticides that are also highly toxic to humans. These insecticides act by blocking the transmission of electrical signals in the respiratory system, thereby causing paralysis and death by suffocation. Fortunately, such poisons do not last for long in the environment because they are destroyed over a period of several days by reaction with water. An important example of an organophosphorus insecticide is malathion, which was used to combat the Mediterranean fruit fly infestation in California in the summer of 1981. Malathion is toxic to humans, but only at fairly large doses. There is an enzyme in human gastric juice that decomposes malathion (insects lack this enzyme). Thus malathion is most toxic to humans when it is absorbed directly into the bloodstream, as, for example, when it comes into contact with a cut in the skin.

$$\begin{array}{l} CH_3O \\ \quad \quad S{=}P{-}S{-}CHCOOCH_2CH_3 \\ CH_3O \quad \quad \quad \quad | \\ \quad \quad \quad \quad \quad CH_2COOCH_2CH_3 \end{array}$$

Malathion

Some commercially important compounds of phosphorus are given in Table 7-8.

Table 7-8 Some important compounds of phosphorus

Compound	*Uses*
phosphorus(V) sulfide, P_2S_5	safety matches, oil additive
phosphorus(V) oxide, P_4O_{10}	dehydrating agent
phosphoric acid, H_3PO_4	fertilizers, soaps and detergents, soft drinks, rust-proofing metals, soil stabilizer
sodium phosphates	synthetic detergents, water softeners, leavening agents
calcium phosphates, $CaHPO_4$ and $Ca(H_2PO_4)_2$	fertilizers, poultry and animal feeds

7-13 ARSENIC AND ANTIMONY ARE SEMIMETALS

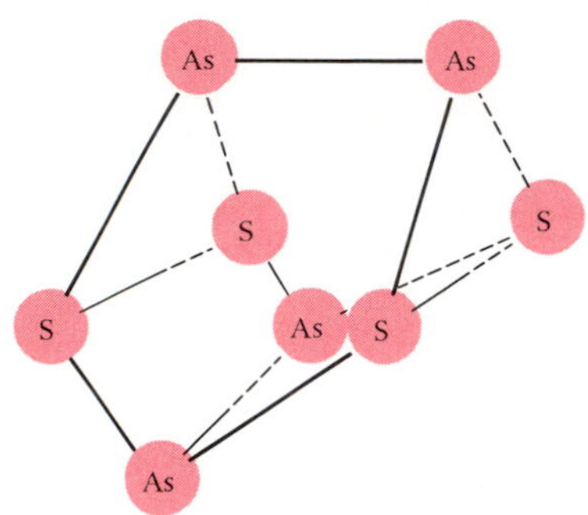

Structure of As_4S_4.

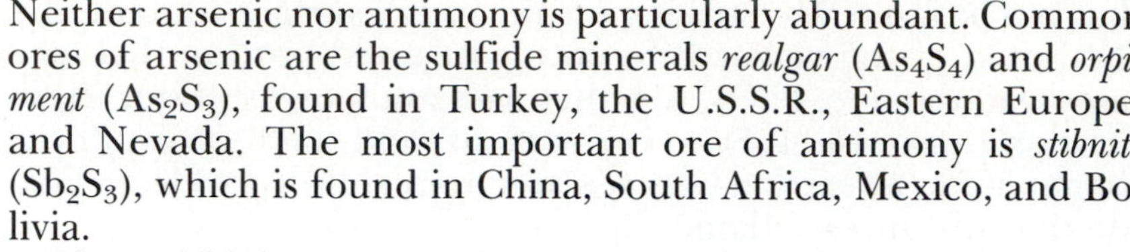

Neither arsenic nor antimony is particularly abundant. Common ores of arsenic are the sulfide minerals *realgar* (As_4S_4) and *orpiment* (As_2S_3), found in Turkey, the U.S.S.R., Eastern Europe, and Nevada. The most important ore of antimony is *stibnite* (Sb_2S_3), which is found in China, South Africa, Mexico, and Bolivia.

The sulfides are converted to the oxides by roasting in air:

$$2As_2S_3(s) + 9O_2(g) \rightarrow 2As_2O_3(s) + 6SO_2(g)$$

The oxides are reduced to the elements with carbon or hydrogen:

$$2As_2O_3(s) + 3C \rightarrow 3CO_2 + 4As(l)$$

Both arsenic and antimony react directly with the halogens to form trihalides and pentahalides such as AsF_3 and SbF_5.

Like other fourth-row post-transition elements, arsenic has a tendency to favor an oxidation state of two less than the maximum for the group. Thus, although PCl_5 and $SbCl_5$ are stable species, $AsCl_5$ decomposes above $-50°C$. Furthermore, antimony trioxide can be prepared by burning antimony in oxygen, but the pentoxide cannot be produced this way. The relative instability of the +5 oxidation state of arsenic means that As_4O_{10} and H_3AsO_4 are strong oxidizing agents.

In conformity with the fact that oxides tend from acidic to basic down any one group in the periodic table, the oxides of arsenic are acidic:

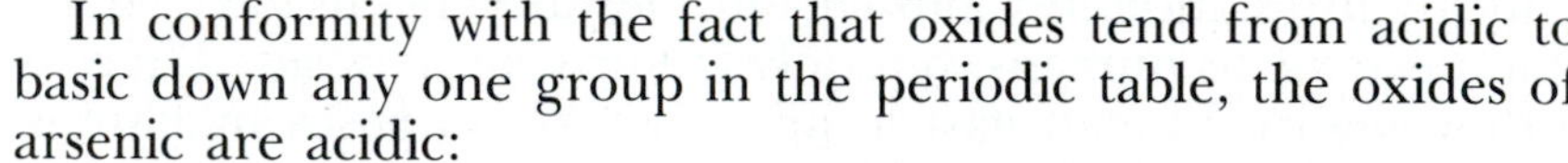

$$As_4O_6(s) + 12NaOH(aq) \rightarrow \underset{\text{sodium arsenite}}{4Na_3AsO_3(aq)} + 6H_2O(l)$$

$$As_4O_{10}(s) + 12NaOH(aq) \rightarrow \underset{\text{sodium arsenate}}{4Na_3AsO_4(aq)} + 6H_2O(l)$$

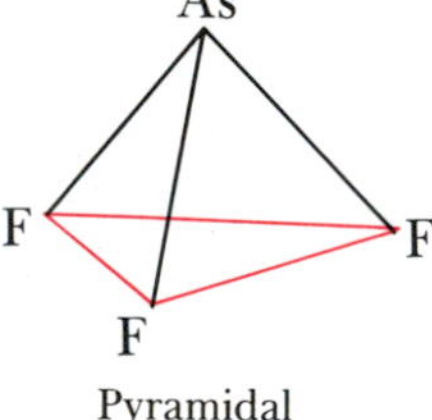

Pyramidal

Arsenic trifluoride has a trigonal pyramidal structure in the gas stage.

and the oxides of antimony are amphoteric:

$$2Sb_2O_3(s) + 6HCl(aq) \rightarrow SbCl_3(s) + 3H_2O(l)$$

$$Sb_2O_3(s) + 6NaOH(aq) \rightarrow \underset{\text{sodium antimonite}}{2Na_3SbO_3(aq)} + 3H_2O(l)$$

and bismuth(III) oxide is basic:

$$Bi_2O_3(s) + 6HCl(aq) \rightarrow 2BiCl_3(s) + 3H_2O(l)$$

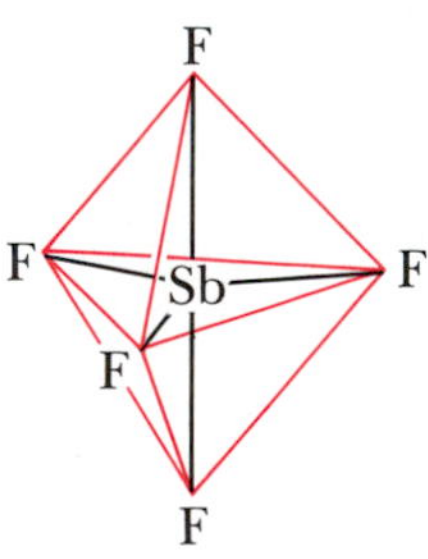

Trigonal bipyramidal

Antimony pentafluoride has a trigonal bipyramidal structure in the gas stage.

Although arsenic compounds are poisonous, trace amounts of arsenic are essential to the growth of red blood cells in bone marrow, and the average healthy human body contains about 7 milligrams of arsenic.

Antimony is distinctly more metallic in character than arsenic, and numerous alloys contain antimony, which acts to prevent

The minerals orpiment (left), stibnite (center), and realgar (right).

corrosion and to increase the resistance of the alloys to fracturing as a result of thermal shock. Some compounds of arsenic and antimony and their uses are given in Table 7-9.

7-14 BISMUTH IS THE ONLY GROUP 5 METAL

The principal compounds of bismuth contain Bi(III), which is two less than the group number. Bismuth is a pink-white metal

Table 7-9 Some compounds of arsenic and antimony

Compound	*Uses*
lead(II) arsenate, $Pb_3(AsO_4)_2(s)$	insecticide
sodium arsenite, $NaAsO_2(s)$	arsenical soap, antiseptic, herbicide, insecticide, fungicide
arsenic(III) oxide, $As_4O_6(s)$	manufacture of glass, insecticide, rodenticide, wood preservative, preparation of many arsenic compounds
antimony trioxide, $Sb_2O_3(s)$	white paint pigment, ceramic opacifier, flame-proofing agent
antimony trichloride, $SbCl_3(s)$	dye mordant, fire-proofing textiles
antimony sulfide, $Sb_2S_3(s)$	vermilion or yellow pigment, pyrotechnics, ruby glass

Table 7-10 Some compounds of bismuth

Compound	*Uses*
bismuth oxychloride, $BiOCl(s)$	face powder, artificial pearls
bismuth subcarbonate $(BiO)_2CO_3(s)$	ceramic glazes, pharmacology
bismuth subnitrate $BiONO_3(s)$	cosmetics, enamel flux

that occurs rarely as the free metal. The most common source of bismuth is the sulfide ore *bismuthinite,* Bi_2S_3.

Bismuth metal is obtained from the ore by roasting it with carbon in air. Bismuth is also obtained as a by-product in lead smelting. Bismuth is used in a variety of alloys, including pewter and low-melting alloys that are used in fire-extinguisher sprinkler-head plugs, electrical fuses, and relief valves for compressed-gas cylinders. Bismuth alloys contract on heating and thus find use in alloys that might otherwise crack because of thermal expansion when subjected to high temperatures.

The oxide Bi_2O_3 is soluble in strongly acidic aqueous solutions (Figure 7-12). The bismuthyl ion, $BiO^+(aq)$, and the bismuthate ion, $BiO_3^-(aq)$, are important in the aqueous-solution chemistry of bismuth. The bismuthyl ion forms insoluble compounds such as BiOCl and BiO(OH), whereas BiO_3^- is a powerful oxidizing agent. Bismuth pentafluoride, BiF_5, is a potent fluorinating agent that transfers fluorine to various compounds and is converted to the trifluoride, BiF_3.

Some bismuth compounds and their major uses are given in Table 7-10.

TERMS YOU SHOULD KNOW

phosphate rock
nitrogen fixation
Haber process
Ostwald process
Raschig synthesis
photodissociation
amorphous
superphosphate
eutrophication

Figure 7-12 Bismuth metal and bismuth(III) oxide, Bi_2O_3.

QUESTIONS

7-1. Complete and balance the following equations.

(a) $P_4(s) + O_2(g) \rightarrow$
excess

(b) $P_4O_6(s) + H_2O(l) \rightarrow$

(c) $P_4O_{10}(s) + H_2O(l) \rightarrow$

(d) $NaN_3(s) \xrightarrow[\text{heat}]{}$

7-2. Complete and balance the following equations.

(a) $N_2O_3(g) + H_2O(l) \rightarrow$

(b) $N_2O_5(s) + H_2O(l) \rightarrow$

(c) $NH_4NO_3(s) \xrightarrow[\text{heat}]{}$

(d) $NH_4NO_2(aq) \xrightarrow[\text{heat}]{}$

7-3. How could you prepare ammonia starting with air, water, and coke?

7-4. Using D_2O as a source of deuterium, how could you prepare $ND_3(g)$?

7-5. Write a chemical equation for the preparation of hydrazoic acid.

7-6. Write the chemical equation for the preparation of hydrazine by the Raschig process.

7-7. Why do solutions of nitric acid often have a brown-yellow color?

7-8. Why are there no nitrogen pentahalides?

7-9. Write Lewis formulas for the following species (include resonance forms where appropriate).

(a) N_2O_4

(b) N_2O_3

(c) CN_2^{2-}

(d) H_2N_2

(e) N_3^-

7-10. Write Lewis formulas for the following species (include resonance forms where appropriate).

(a) $HONH_2$

(b) NO_3^-

(c) NO_2^-

(d) N_2O

(e) NO_2

7-11. How could you prepare DCl, using D_2O as a source of deuterium?

7-12. How could you prepare PD_3, using D_2O as a source of deuterium?

7-13. Write chemical equations for the reactions of phosphorus pentoxide with sulfuric acid and with nitric acid.

7-14. Phosphine dissolves in liquid ammonia to give $NH_4^+PH_2^-$. Use VSEPR theory to predict the shape of PH_2^-.

7-15. Use VSEPR theory to predict the shapes of (a) $POCl_3$, (b) PO_4^{3-}, and (c) PCl_6^-.

7-16. When P_4O_{10} and P_4S_{10} are heated in the appropriate proportions above 400°C, $P_4O_6S_4$, a colorless, hygroscopic crystalline substance, is obtained. Using Figure 7-8 as a guide, predict the structure of $P_4O_6S_4$. What about the structure of $P_4O_4S_6$?

7-17. Write chemical equations describing the acidic characters of As_4O_6 and As_4O_{10}, the amphoteric character of Sb_2O_3, and the basic character of Bi_2O_3.

7-18. Use Lewis formulas to show that H_3PO_2 is monobasic, H_3PO_3 is dibasic, and H_3PO_4 is tribasic.

7-19. Predict which of the following molecular halides hydrolyze, and write a balanced equation for those that do.

(a) SF_4

(b) CCl_4

(c) XeF_6

(d) BrF_5

(e) $AsBr_3$

7-20. Use VSEPR theory to predict the structures of the following species.

(a) AsO_3^{3-}

(b) AsO_4^{3-}

(c) $SbCl_3$

(d) $BiCl_3$

(e) AsF_5

7-21. Suggest why the H—X—H bond angles in NH_3, PH_3, AsH_3, and SbH_3 decrease from 107° for NH_3 to 93° for PH_3, 92° for AsH_3, and 91° for SbH_3.

7-22. For the equilibrium

$$\underset{\text{brown}}{2NO_2(g)} \rightleftharpoons \underset{\text{colorless}}{N_2O_4(g)} \qquad \Delta H^\circ_{rxn} = -57.2 \text{ kJ}$$

does the reaction mixture become increasingly colored with increasing or decreasing temperature?

7-23. Deuterium chloride gas, DCl(*g*), is prepared in the laboratory by the reaction

$$PCl_3(l) + D_2O(l) \rightarrow DCl(g) + D_3PO_3(l)$$

(a) Balance the equation and compute the number of grams of $D_2O(l)$ required to prepare 5.0 mg of DCl(*g*).

(b) Assume that the gas is collected in a 4.0-mL vessel at 20°C and compute the pressure of DCl(*g*) in the vessel.

7-24. Use data in Table 7-6 to compute the pH of the following solutions at 25°C.

(a) 0.10 M $H_3PO_4(aq)$

(b) 0.10 M $H_3PO_3(aq)$

(c) 0.10 M $H_3PO_2(aq)$

THE GROUP 6 ELEMENTS

Although steel wool does not burn in air, it burns vigorously in pure oxygen.

The Group 6 elements are oxygen, sulfur, selenium, tellurium, and polonium. There is a continuous progression in Group 6 from nonmetallic to metallic properties with increasing atomic number. Oxygen, sulfur, and selenium are nonmetals, tellurium is a semimetal, and polonium is a metal. As in other groups, there is a significant difference between the chemical properties of the first member and the second member. Oxygen is limited to two bonds (e.g., H_2O) or occasionally three bonds (e.g., H_3O^+), whereas the other members of the group may uti-

Table 8-1 Sources and uses of the Group 6 elements

Element	*Principal sources*	*Uses*
oxygen	fractional distillation of liquid air	blast furnaces; steel production; production of methyl alcohol, acetylene, ethylene oxide, etc.; rocket propellant; sewage treatment; breathing apparatus; many other uses
sulfur	native from underground deposits (Frasch process); natural gas and petroleum by-product	manufacture of sulfuric acid, paper manufacture, rubber vulcanization, drugs and pharmaceuticals, dyes, fungicides, insecticides
selenium	anode muds of the electrolyte refining of copper and lead	photocells, exposure meters, solar cells, xerography, production of ruby-colored glasses and enamels
tellurium	anode muds of the electrolyte refining of copper and lead	alloys to improve machinability of copper and stainless steels, ceramics

lize *d* orbitals to form compounds such as SF_6 and TeF_6. As in Groups 4 and 5, there is a decrease in thermal stability of the binary hydrogen compounds in going from H_2S to H_2Po. In addition, there is an increasing stability with increasing atomic number of an oxidation state two less than the group number.

Oxygen is the most abundant element on earth and the third most abundant element in the universe, ranking behind hydrogen and helium. Most rocks contain a large amount of combined oxygen. For example, sand is predominantly silicon dioxide (SiO_2) and consists of more than 50 percent oxygen by mass. Almost 90 percent of the mass of the oceans and two thirds of the mass of the human body are oxygen. Air is 21 percent oxygen by volume. We can live weeks without food, days without water, but only minutes without oxygen.

Sulfur is widely distributed in nature, but not usually in sufficient concentration to merit commercial mining. The two most important sources of sulfur are hydrogen sulfide from natural gas and petroleum refining, and elemental sulfur from large salt domes offshore along the Gulf of Mexico. Selenium and tellurium are relatively rare elements. Both elements are found in association with metal sulfide ores, and are obtained commercially as by-products of the refining of copper and lead (Table 8-1). Polonium has no stable isotopes; minute quantities of polonium-210 occur in uranium ores.

▪ Marie Curie was awarded the Nobel Prize in Chemistry in 1911 for the isolation and identification of the element polonium. The element was named in honor of her native country, Poland.

Tables 8-2 and 8-3 present some atomic and physical properties of the Group 6 elements. The usual trends with increasing atomic number are evident.

Table 8-2 Atomic properties of the Group 6 elements

Property	*O*	*S*	*Se*	*Te*	*Po*
atomic number	8	16	34	52	84
atomic mass/amu	15.9994	32.06	78.96	127.60	(209)
number of naturally occurring isotopes	3	4	6	7	0
ground-state electron configuration	$[He]2s^22p^4$	$[Ne]3s^23p^4$	$[Ar]3d^{10}4s^24p^4$	$[Kr]4d^{10}5s^25p^4$	$[Xe]4f^{14}5d^{10}6s^26p^4$
atomic radius/pm	60	100	115	140	190
ionization energy/$MJ \cdot mol^{-1}$					
first	1.31	1.00	0.941	0.869	0.812
second	3.39	2.25	2.05	1.79	—
Pauling electronegativity	3.5	2.5	2.4	2.1	2.0

8-1 THIRTY-FIVE BILLION POUNDS OF OXYGEN ARE SOLD ANNUALLY IN THE UNITED STATES

Oxygen is a colorless, odorless, tasteless gas that exists as a diatomic molecule, O_2. Although colorless in the gaseous state both liquid and solid oxygen are pale blue.

Industrially, oxygen is produced by the fractional distillation of liquid air. Approximately 35 billion pounds of oxygen are sold annually in the United States, making it the fifth most important industrial chemical. The major commercial use of oxygen is in the blast furnaces used to manufacture steel. Oxygen is also used in hospitals, in oxyhydrogen and oxyacetylene torches for welding metals, and to facilitate breathing at high altitudes and under water. Tremendous quantities of oxygen are used directly from air as a reactant in the combustion of hydrocarbon fuels, which supply 93 percent of the energy consumed in the United States. In terms of total usage (pure oxygen and oxygen used directly from air), oxygen is the number two chemical, ranking behind only water.

Liquid oxygen.

8-2 OXYGEN IN THE EARTH'S ATMOSPHERE IS PRODUCED BY PHOTOSYNTHESIS

Most of the oxygen in the atmosphere is the result of *photosynthesis,* the process by which green plants combine $CO_2(g)$ and $H_2O(g)$ into carbohydrates and $O_2(g)$ under the influence of visi-

Table 8-3 Some physical properties of the Group 6 elements

Property	O_2	*S*	*Se*	*Te*
melting point/°C	−218.8	115	221	450
boiling point/°C	−183.0	445	685	1009
$\Delta\overline{H}_{fus}/kJ \cdot mol^{-1}$	0.443	1.72	5.44	17.5
$\Delta\overline{H}_{vap}/kJ \cdot mol^{-1}$	6.82	8.37		50.6
density at 20°C/$g \cdot cm^{-3}$	1.33×10^{-3}	1.96	4.79	6.24

ble light. The carbohydrates appear in the plants as starch, cellulose, and sugars. The reaction is described schematically as

$$CO_2(g) + H_2O(g) \xrightarrow{\text{visible light}} \text{carbohydrate} + O_2(g)$$

When the carbohydrate is glucose, we have

$$6CO_2(g) + 6H_2O(l) \rightarrow C_6H_{12}O_6(aq) + 6O_2(g)$$
$$\Delta G^\circ_{rxn} = +2870 \text{ kJ at } 25°C$$

The reaction is driven up the Gibbs free energy "hill" ($\Delta G^\circ_{rxn} >> 0$) by the energy obtained from sunlight.

▪ The oxygen evolved in photosynthesis reactions arises from the oxygen in the water molecules, as shown by studies with the heavy isotope, ^{18}O. That is, if the photosynthesis reaction is carried out with $H_2{}^{18}O$ water, then the evolved oxygen is $^{18}O_2$, whereas with $C^{18}O_2$, the evolved O_2 has the normal isotopic composition.

In one year, more than 10^{10} metric tons of carbon is incorporated into carbohydrates by photosynthesis. In the hundreds of millions of years that plant life has existed on earth, photosynthesis has produced much more oxygen than the amount now present in the atmosphere.

8-3 OXYGEN CAN BE PREPARED IN THE LABORATORY BY DECOMPOSITION REACTIONS

A frequently used method for preparing oxygen in the laboratory is the thermal decomposition of potassium chlorate, $KClO_3$. The chemical equation for the reaction is

$$2KClO_3(s) \xrightarrow{MnO_2(s)} 2KCl(s) + 3O_2(g)$$

This reaction requires a temperature of about 400°C, but if a small amount of the catalyst manganese dioxide, MnO_2, is added, the reaction occurs rapidly at 250°C. An alternate method for the laboratory preparation of oxygen is to add sodium peroxide, Na_2O_2, to water:

$$2Na_2O_2(s) + 2H_2O(l) \rightarrow 4NaOH(aq) + O_2(g)$$

This rapid and convenient reaction does not require heat. However, oxygen also can be prepared by the electrolysis of water (Chapter 2):

$$2H_2O(l) \xrightarrow{\text{electrolysis}} 2H_2(g) + O_2(g)$$

8-4 OXYGEN REACTS DIRECTLY WITH MOST OTHER ELEMENTS

Oxygen is very reactive, and is the second most electronegative element. It reacts directly with all the other elements except the halogens, the noble gases, and some of the less reactive metals to form a wide variety of compounds. Only fluorine reacts with more elements than oxygen. Compounds containing oxygen constitute 31 of the top 50 industrial chemicals (Appendix A).

Oxygen forms oxides with many elements. Most metals react rather slowly with oxygen at ordinary temperatures but react more rapidly as the temperature is increased. For example, iron, in the form of steel wool, burns vigorously in pure oxygen but does not burn in air (see frontispiece).

Methane, the main constituent of natural gas, burns in oxygen according to the equation

$$\underset{\text{methane}}{CH_4(g)} + 2O_2(g) \rightarrow CO_2(g) + 2H_2O(g) \qquad \Delta H^\circ_{rxn} = -802 \text{ kJ}$$

All hydrocarbons burn in oxygen to give carbon dioxide and water. Gasoline is a mixture of hydrocarbons. Using octane, C_8H_{18}, as a typical hydrocarbon in gasoline, we write the combustion of gasoline as

$$\underset{\text{octane}}{2C_8H_{18}(l)} + 25O_2(g) \rightarrow 16CO_2(g) + 18H_2O(g) \qquad \Delta H^\circ_{rxn} = -5460 \text{ kJ}$$

The energy released in reactions of this type is used to power machinery and to produce electricity (Chapter 6 of the text).

A mixture of acetylene and oxygen is burned in the oxyacetylene torch. The chemical equation for the combustion of acetylene is

▪ Different flames have different temperatures. The temperature of a hot region of a candle flame is about 1200°C, that of a Bunsen burner flame is about 1800°C, and that of the flame of an oxyacetylene torch is about 2400°C.

$$\underset{\text{acetylene}}{2C_2H_2(g)} + 5O_2(g) \rightarrow 4CO_2(g) + 2H_2O(g)$$

The flame temperature of an oxyacetylene welding torch is about 2400°C, which is sufficient to melt iron and steel. A combustion reaction with which we are all familiar is the burning of a candle. The wax in a candle is composed of long-chain hydrocarbons, such as $C_{20}H_{42}$. The molten wax rises up the wick to the combustion zone the way ink rises in a piece of blotting paper.

8-5 SOME METALS REACT WITH OXYGEN TO YIELD PEROXIDES

Although most metals yield oxides when they react with oxygen, some of the more reactive metals, such as sodium, potassium, and rubidium, yield peroxides and superoxides. Peroxides are compounds in which the negative ion is the peroxide ion, O_2^{2-}, and superoxides are compounds in which the negative ion is the superoxide ion, O_2^-. For example,

$$2Na(s) + O_2(g) \rightarrow \underset{\text{sodium peroxide}}{Na_2O_2(s)}$$

$$K(s) + O_2(g) \rightarrow \underset{\text{potassium superoxide}}{KO_2(s)}$$

▪ The ground state of the oxygen molecule is paramagnetic.

Table 8-4 Comparison of diatomic oxygen species

Species	*Molecular-orbital valence electron occupancy*	*Net number of bonding electrons*	*Bond length*/pm	*Bond energy*/kJ
O_2^+	$(2\sigma)^2(2\sigma^*)^2(3\sigma)^2(1\pi)^4(1\pi^*)^1$	5	112	640
O_2	$(2\sigma)^2(2\sigma^*)^2(3\sigma)^2(1\pi)^4(1\pi^*)^2$	4	121	506
O_2^-	$(2\sigma)^2(2\sigma^*)^2(3\sigma)^2(1\pi)^4(1\pi^*)^3$	3	126	370
O_2^{2-}	$(2\sigma)^2(2\sigma^*)^2(3\sigma)^2(1\pi)^4(1\pi^*)^4$	2	149	160

Table 8-4 gives a comparison of O_2 with the ions O_2^+, O_2^-, and O_2^{2-}.

One of the most important peroxides is hydrogen peroxide, H_2O_2, a colorless, syrupy liquid that explodes violently when heated. Hydrogen peroxide is a strong oxidizing agent that oxidizes a wide variety of organic substances. Dilute aqueous solutions of hydrogen peroxide are fairly safe to use. A 3% aqueous solution is sold in drugstores and used as a mild antiseptic and as a bleach. More concentrated solutions (30%) of hydrogen peroxide are used industrially as a bleaching agent for hair, flour, textile fibers, fats, and oils, in the artificial aging of wines and liquor, and for control of pollution in sewage effluents. Hydrogen peroxide can act as either an oxidizing agent (electron acceptor):

Table 8-5 Commercially available aqueous solutions of hydrogen peroxide

H_2O_2 concentration	*Use*
3%	antiseptic
6%	hair bleach
30%	industrial and laboratory oxidizing agent
85%	potent oxidizing agent

(1) $$2H^+(aq) + H_2O_2(aq) + 2e^- \rightarrow 2H_2O(l)$$

or as a reducing agent (electron donor):

(2) $$H_2O_2(aq) \rightarrow O_2(g) + 2H^+(aq) + 2e^-$$

Note that the sum of half-reactions (1) and (2) is

$$2H_2O_2(aq) \rightarrow 2H_2O(l) + O_2(g)$$

or in other words, hydrogen peroxide self-destructs via a disproportionation (self–oxidation-reduction) reaction.

8-6 OZONE, O_3, IS A POTENT OXIDIZING AGENT

When a spark is passed through oxygen, some of the oxygen is converted to ozone, O_3:

$$3O_2(g) \rightarrow 2O_3(g)$$

Ozone is a light blue gas at room temperature. It has a sharp, characteristic odor that occurs after electrical storms or near high-voltage generators. Liquid ozone (boiling point -112°C) is a deep blue, explosive liquid. Ozone is so reactive that it cannot be transported safely, but must be generated as needed. Relatively unreactive metals, such as silver and mercury, which do not react with oxygen, react with ozone to form oxides. Ozone is used as a bleaching agent and is being considered as a replacement for chlorine in water treatment because of the environmental problem involving chlorinated hydrocarbons. Ozone is produced in the stratosphere by the reactions

■ Ozone is one of the strongest known oxidizing agents.

$$O_2(g) \xrightarrow{\lambda < 240 \text{ nm}} 2O(g)$$

$$O(g) + O_2(g) \rightarrow O_3(g)$$

Stratospheric ozone screens out ultraviolet light in the wavelength region 240 to 310 nm via the reaction

$$O_3(g) \xrightarrow{h\nu} O_2(g) + O(g)$$

and thereby protects life on earth from destruction by such light.

8-7 SULFUR OCCURS AS THE FREE ELEMENT

Sulfur is a yellow, tasteless, odorless solid that is often found in nature as the free element. Sulfur is essentially insoluble in water but dissolves readily in carbon disulfide, CS_2. It does not react with dilute acids or bases, but it does react with many metals at elevated temperatures to form metal sulfides.

Sulfur, which constitutes only 0.05 percent of the earth's crust, is not one of the most prevalent elements. Yet it is one of the most commercially important ones because it is the starting material for the most important industrial chemical, sulfuric acid.

Prior to 1900, most of the world's supply of sulfur came from Sicily, where sulfur occurs at the surfaces around hot springs and volcanoes. In the early 1900s, however, large subsurface deposits of sulfur were found along the Gulf Coast of the United States. The sulfur occurs in limestone caves, over 1000 feet beneath layers of rock, clay, and quicksand. The recovery of the sulfur from these deposits posed a great technological problem, which was solved by the engineer Herman Frasch. The *Frasch process* (Figure 8-1) uses an arrangement of three concentric pipes (diameters of 1 in., 3 in., and 6 in.) placed in a bore hole that penetrates to the base of the sulfur-bearing calcite ($CaCO_3$) rock formation. Pressurized hot water (180°C) is forced down the space between the 6-in. and 3-in. pipes to melt the sulfur (melting point 119°C). The molten sulfur, which is twice as dense as water, sinks to the bottom of the deposit and is then forced up the space between the 3-in. and 1-in. pipes as a foam by the action of compressed air injected through the innermost pipe. The molten sulfur rises to the surface, where it is pumped into

Sulfur with heated hot water and air can be seen surfacing from Culberson Mine in West Texas.

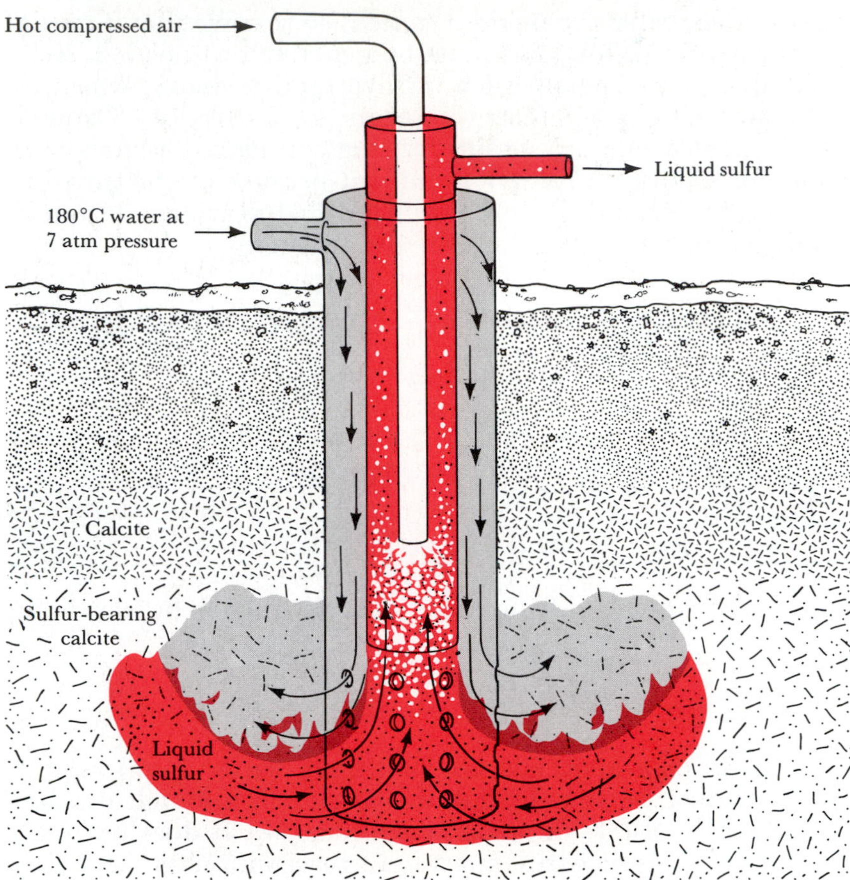

Figure 8-1 The Frasch process for sulfur extraction. Three concentric pipes are sunk into sulfur-bearing calcite rock. Water at 180°C is forced down the outermost pipe to melt the sulfur. Hot compressed air is forced down the innermost pipe and mixes with the molten sulfur, forming a foam of water, air, and sulfur. The mixture rises to the surface through the center pipe. The resulting dried sulfur has a purity of 99.5 percent.

tank cars for shipment or into storage areas. A significant part of the U.S. annual sulfur production of over 10 million metric tons is obtained by the Frasch process from the region around the Gulf of Mexico in Louisiana and Texas (Figure 8-2).

Sulfur is also obtained in increasingly large quantities from the hydrogen sulfide (H_2S) in so-called sour natural gas and from H_2S produced when sulfur is removed from petroleum. Hydrogen sulfide is burned in air to produce sulfur dioxide gas, which is then reacted with additional hydrogen sulfide to produce sulfur:

$$2H_2S(g) + 3O_2(g) \rightarrow 2SO_2(g) + 2H_2O(g)$$

$$SO_2(g) + 2H_2S(g) \rightarrow 3S(l) + 2H_2O(g)$$

These reactions are also thought to be responsible for the surface deposits of sulfur around hot springs and volcanoes.

Texasgulf

Figure 8-2 Sulfur is mined in by the Frasch process. It is then formed into huge blocks ready for shipment such as these in Newhall, Texas.

8-8 SULFIDE ORES ARE IMPORTANT SOURCES OF SEVERAL METALS

Deposits of metal sulfides are found in many regions and are valuable ores of the respective metals. Galena (PbS), cinnabar (HgS), and antimony sulfide (Sb_2S_3) are examples of metal sulfides that are ores. In obtaining metals from sulfide ores, the ores are usually *roasted,* meaning that they are heated in an oxygen atmosphere. The roasting of galena is described in Section 6-9.

Iron pyrite, also known as fool's gold, is a famous metal sulfide that has little commercial value. Sulfur is also found in nature in a few insoluble sulfates, such as *gypsum,* $CaSO_4 \cdot 2H_2O$ (calcium sulfate dihydrate) (Figure 8-3), and *barite,* $BaSO_4$.

Sulfur also occurs in many proteins. Hair protein is fairly rich in sulfur. In fact, the formation of a "permanent" wave in hair involves the breaking and remaking of sulfur bonds.

8-9 SULFUR EXISTS AS RINGS OF EIGHT SULFUR ATOMS

Below 96°C sulfur exists as yellow, transparent, *rhombic* crystals, shown in Figure 8-4(a). If rhombic sulfur is heated above 96°C, then it becomes opaque and the crystals expand into *monoclinic* crystals (Figure 8-4(b)). Monoclinic sulfur is the stable form from 96°C to the melting point. The molecular units of the rhombic form are rings containing eight sulfur atoms, S_8 (Figure 8-5). The molecular units of monoclinic sulfur are also S_8 rings, but the rings themselves are arranged differently.

▪ Cold rhombic sulfur is colorless.

Monoclinic sulfur melts at 119°C to a thin, pale yellow liquid consisting of S_8 rings. Upon heating to about 150°C there is little

Alain Pitcairn/Grant Heilman

Ron Testa/Field Museum

Figure 8-3 Large deposits of gypsum, $CaSO_4 \cdot 2H_2O$, an insoluble mineral, are found in many areas. Left, the dunes of White Sands National Monument in New Mexico are composed of gypsum. Above is a 3-inch cluster of gypsum crystals.

change, but beyond 150°C the liquid sulfur begins to thicken and turns reddish brown. By 200°C, the liquid is so thick that it hardly pours (Figure 8-6). The molecular explanation for this behavior is simple. At about 150°C, thermal agitation causes the S_8 rings to begin to break apart and form chains of sulfur atoms:

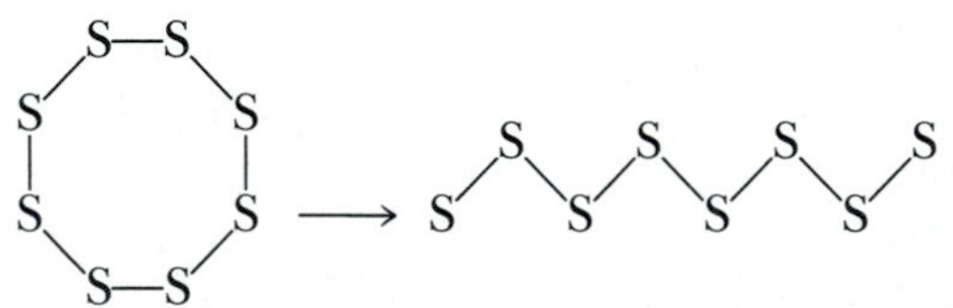

Smithsonian

(a)

Photo Researchers

(b)

Figure 8-4 Sulfur occurs as (a) rhombic and (b) monoclinic crystals. Rhombic sulfur is the stable form below 96°C. From 96°C to 119°C (the normal melting point) monoclinic sulfur is the stable form. The terms rhombic and monoclinic are derived from the shape of the crystals.

These chains can then join together to form longer chains, which become entangled in each other and cause the liquid to thicken. Above 250°C, the liquid begins to flow more easily because the thermal agitation is sufficient to begin to break the chains of sulfur atoms. At the boiling point (445°C), liquid sulfur pours freely and the vapor molecules consist mostly of S_8 rings.

If liquid sulfur at about 200°C is placed quickly in cold water (this process is called *quenching*), then a rubbery substance known as plastic sulfur is formed. The material is rubbery because the long, coiled chains of sulfur atoms can straighten out some if they are pulled. As plastic sulfur cools, it slowly becomes hard again as it rearranges itself into the rhombic form.

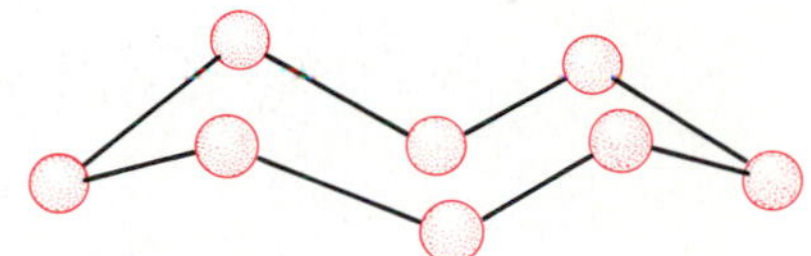

Figure 8-5 Under most conditions, sulfur exists as eight-membered rings, S_8. The ring is not flat but puckered in such a way that four of the atoms lie in one plane and the other four lie in another plane.

8-10 SULFURIC ACID IS THE LEADING INDUSTRIAL CHEMICAL

By far the most important use of sulfur is in the manufacture of sulfuric acid. Most sulfuric acid is made by the *contact process*. The sulfur is first burned in oxygen to produce sulfur dioxide:

$$S(s) + O_2(g) \rightarrow SO_2(g)$$

The sulfur dioxide is then converted to sulfur trioxide in the presence of the catalyst vanadium pentoxide:

$$2SO_2(g) + O_2(g) \xrightarrow{V_2O_5(s)} 2SO_3(g)$$

Figure 8-6 Molten sulfur at various temperatures. The change in color and physical properties of liquid sulfur with increasing temperature (120° to about 250°C) is a result of the conversion of eight-membered rings to long chains of sulfur atoms. Above 250°C, the long chains begin to break up into smaller segments and the sulfur is more fluid.

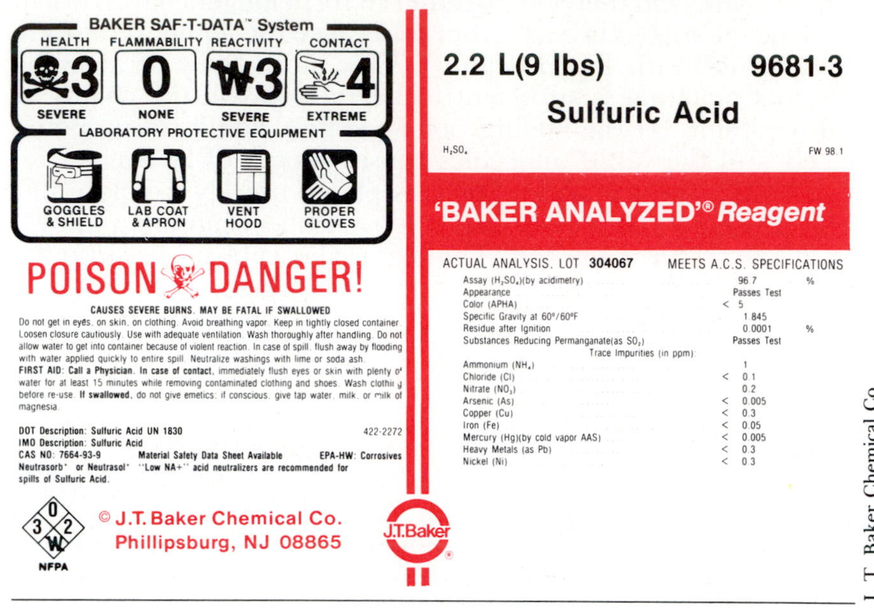

J. T. Baker Chemical Co.

Figure 8-7 Sulfuric acid is sold for laboratory use as an 18 M solution that is 98 percent sulfuric acid and 2 percent water.

The sulfur trioxide is then absorbed into nearly pure liquid sulfuric acid to form *fuming sulfuric acid (oleum):*

$$H_2SO_4(l) + SO_3(g) \rightarrow \underset{\text{oleum}}{H_2S_2O_7}(35\% \text{ in } H_2SO_4)$$

The oleum is then added to water or aqueous sulfuric acid to produce the desired final concentration of aqueous sulfuric acid. Sulfur trioxide cannot be absorbed directly in water because the acid mist of H_2SO_4 that forms is very difficult to condense.

Over 60 billion pounds of sulfuric acid are produced annually in the United States. Commercial-grade sulfuric acid is one of the least expensive chemicals, costing less than 10 cents per pound in bulk quantities. Very large quantities of sulfuric acid are used in the production of fertilizers and numerous industrial chemicals, the petroleum industry, metallurgical processes and synthetic fiber production.

Pure, anhydrous sulfuric acid is a colorless, syrupy liquid that freezes at 10°C and boils at 290°C. The standard laboratory acid is 98 percent H_2SO_4 and 18 M in H_2SO_4 (Figure 8-7). Concentrated sulfuric acid is a powerful dehydrating agent. Gases are sometimes bubbled through it to remove traces of water vapor—provided, of course, that the gases do not react with the acid.

Sulfuric acid is such a strong dehydrating agent that it can remove water from carbohydrates, such as cellulose and sugar, even though these substances contain no free water. If concentrated sulfuric acid is poured over sucrose, $C_{12}H_{22}O_{11}$, then we have the reaction

$$C_{12}H_{22}O_{11}(s) \xrightarrow[H_2SO_4\ (98\%)]{} 12C(s) + 11H_2O \text{ (in } H_2SO_4)$$

Figure 8-8 Concentrated (98%) sulfuric acid is a powerful dehydrating agent capable of converting sucrose to carbon.

This impressive reaction is shown in Figure 8-8. Similar reactions are responsible for the destructive action of concentrated sulfuric acid on wood, paper, and skin.

The high boiling point and strength of sulfuric acid are the basis of its use in the production of other acids. For example, dry hydrogen chloride gas is produced by the reaction of sodium chloride with sulfuric acid:

$$2NaCl(s) + H_2SO_4(l) \rightarrow Na_2SO_4(s) + 2HCl(g)$$

The high boiling point of the sulfuric acid allows the $HCl(g)$ to be driven off by heating. The $HCl(g)$ is then added to water to produce hydrochloric acid. Note that this reaction is a double replacement reaction driven by the removal of a gaseous product from the reaction mixture.

8-11 SULFUR FORMS SEVERAL WIDELY USED COMPOUNDS

Most metal sulfides react with strong acids to produce the foul-smelling, very poisonous gas, hydrogen sulfide. For example,

$$ZnS(s) + H_2SO_4(aq) \rightarrow ZnSO_4(aq) + H_2S(g)$$

Hydrogen sulfide is detectable by smell at low concentrations, but at high concentrations H_2S readily saturates the olfactory sense and the presence of the gas cannot be detected by smell.

Trace amounts of hydrogen sulfide occur naturally in the atmosphere due to volcanic activity and the decay of organic matter. In fact, the presence of hydrogen sulfide in the atmosphere is demonstrated by the tarnishing of silver. In the presence of

Silver sulfide is a black, insoluble solid that appears as a dark tarnish on the surface of silver.

Table 8-6 Some compounds of sulfur

Compound	*Uses*
sulfuric acid, $H_2SO_4(l)$	manufacture of fertilizers, dyes, explosives, steel, and other acids; petroleum industry; metallurgy; plastics
sulfur dioxide, $SO_2(g)$	disinfectant in the food and brewing industries; bleaching agent for paper, textiles, oils, etc.; fumigant, preservative
aluminum sulfate, $Al_2(SO_4)_3(s)$	leather tanning; sizing paper; fire-proofing and water-proofing cloth; clarifying agent for oils and fats; water treatment; decolorizer and deodorizer; antiperspirants
ammonium sulfate, $(NH_4)_2SO_4(s)$	fertilizer; water treatment; fire-proofing fabrics; tanning; food additive
sodium sulfate, $Na_2SO_4(s)$	manufacture of paper and glass; textiles; dyes; ceramic glazes; pharmaceuticals; solar heat storage (as the decahydrate, called Glauber's salt)
carbon disulfide, $CS_2(l)$	production of rayon, carbon tetrachloride, cellophanes, soil disinfectants; solvent

oxygen, silver reacts with hydrogen sulfide according to the reaction

$$4Ag(s) + 2H_2S(g) + O_2(g) \rightarrow \underset{\text{black}}{2Ag_2S(s)} + 2H_2O(l)$$

Hydrogen sulfide is a very weak diprotic acid in water ($pK_{a1} = 7$, $pK_{a2} = 13$, at 25°C), and thus the sulfide ion $S^{2-}(aq)$ has a high affinity for protons in water. This is the reason why metal sulfides, most of which are insoluble in water, dissolve readily in aqueous solutions of strong acids, as illustrated previously for ZnS. Hydrogen sulfide is an important reagent in various qualitative analysis schemes in which metal ions are selectively removed from solution as insoluble metal sulfides (Section 19-13 of the text).

$$\begin{array}{c} CH_3 \quad NH_2 \\ \diagdown \; \diagup \\ C \\ \| \\ S \end{array}$$

Thioacetamide

The organosulfur compound thioacetamide, CH_3CSNH_2, is now often used in qualitative analysis schemes as a controlled source of hydrogen sulfide, which is generated by heating a solution of thioacetamide:

$$CH_3\underset{\underset{S}{\|}}{C}NH_2(aq) + 2H_2O(l) \xrightarrow{60°C} CH_3COO^-(aq) + NH_4^+(aq) + H_2S(aq)$$

By using thioacetamide it is possible to keep the H_2S content of laboratory air below harmful levels.

Sulfur burns in oxygen to form sulfur dioxide, a colorless gas with a characteristic choking odor. A pressure of 3 atm is sufficient to liquefy sulfur dioxide at 20°C. At one time SO_2 was used in industrial refrigeration units, but the unpleasant odor and toxicity brought on its replacement by Freons.

Most sulfur dioxide is used to make sulfuric acid, but some is used as a bleaching agent in the manufacture of paper products, oils and starch, and as a food additive to inhibit browning. Large quantities are used in the wine industry as a fungicide for grapevines and as an antioxidant for wines.

Sulfur dioxide is very soluble in water; over 200 g of sulfur dioxide dissolve in one liter of water. Some of the sulfur dioxide reacts with the water to form sulfurous acid:

$$SO_2(g) + H_2O(l) \rightarrow H_2SO_3(aq)$$

but most of its exists in solution as $SO_2(aq)$.

The salts of sulfurous acid are called sulfites. For example, if sodium hydroxide is added to an aqueous solution of sulfur dioxide, then sodium sulfite is formed according to the equation

$$2NaOH(aq) + H_2SO_3(aq) \rightarrow \underset{\text{sodium sulfite}}{Na_2SO_3(aq)} + 2H_2O(l)$$

The sulfite ion is a mild reducing agent that is used in the textile and paper industries to destroy excess chlorine, which is used as a bleaching agent:

$$Cl_2(aq) + SO_3^{2-}(aq) + H_2O(l) \rightarrow 2Cl^-(aq) + SO_4^{2-}(aq) + 2H^+(aq)$$

Sodium sulfite is used occasionally as a preservative, especially for dehydrated fruits. The sulfite ion acts as a fungicide; however, it imparts a characteristic sulfur dioxide odor and taste to the food.

The thiosulfate ion is produced when an aqueous solution of a metal sulfite, such as $Na_2SO_3(aq)$, is boiled in the presence of solid sulfur:

▪ The designation thio denotes the replacement of an oxygen atom by a sulfur atom.

$$S(s) + SO_3^{2-}(aq) \rightarrow \underset{\text{thiosulfate}}{S_2O_3^{2-}(aq)}$$

The thiosulfite ion has a tetrahedral structure. Note that the two sulfur atoms in $S_2O_3^{2-}$ are not equivalent. The structure is analogous to that of sulfate ion with one of the oxygen atoms replaced by a sulfur ion. Thiosulfate ion is used extensively as "hypo" ($Na_2S_2O_3 \cdot 5H_2O$) in black-and-white photography (Section 11-6). Thiosulfate ion is a moderately strong reducing agent that reacts with aqueous iodine or triiodide ion (I_3^-) to form tetrathionate ion, $S_4O_6^{2-}(aq)$:

$$2S_2O_3^{2-}(aq) + I_3^-(aq) \rightarrow \underset{\text{tetrathionate}}{S_4O_6^{2-}(aq)} + 3I^-(aq)$$

This reaction is used in the analytical determination of triiodide ion, which is produced by the action of many mild oxidizing agents on $I^-(aq)$.

8-12 OXIDES OF SULFUR ARE MAJOR POLLUTANTS OF THE ATMOSPHERE

Two oxides of sulfur, SO_2 and SO_3, are major atmospheric pollutants in industrial and urban areas. Most coal and petroleum contain some sulfur, which becomes SO_2 when burned. Concentrations of SO_2 as low as 0.1 to 0.2 ppm can be incapacitating to persons suffering from respiratory conditions such as emphysema and asthma. Although SO_2 is not easily oxidized to SO_3, the presence of dust particles and other particulate matter or ultraviolet radiation facilitates the conversion. The SO_3 then reacts with water vapor to form a very fine sulfuric acid mist. Such a mist is also produced in automobile catalytic converters. Both sulfuric acid and sulfurous acid, H_2SO_3, which arises from the reaction

$$SO_2(g) + H_2O(g) \rightarrow H_2SO_3(\text{mist})$$

produce acid rain, which is rain that is up to 1000 times more acidic than normal rain. Acid rain occurs commonly in northern Europe and in the northeastern United States and Canada. Many lakes in these regions are so acidic as a result of acid rain that the fish life is disappearing.

Acid rain has a devastating effect on limestone and marble, both of which contain $CaCO_3$. The reaction that occurs is

$$CaCO_3(s) + H_2SO_4(aq) \rightarrow CaSO_4(s) + H_2O(l) + CO_2(g)$$

The formation of powdered calcium sulfate breaks down the limestone or marble structure. The decomposition of carbonates by acid rain is a major cause of the deterioration of the ancient buildings and monuments of Europe.

There have been three major disasters attributed to air polluted with oxides of sulfur. In 1952, a gray fog highly polluted with oxides of sulfur settled over London for several days and was reportedly responsible for 4000 deaths. Such a *London fog*, as it is now called, also caused hundreds of deaths in Donora, Pennsylvania, in 1948 and along the Meuse Valley in Belgium in 1930.

Several methods can be used to control the amount of SO_2 introduced into the atmosphere. One obvious way is to burn low-sulfur coal and petroleum. Nigerian oil and some Middle East oil is low in sulfur, whereas Venezuelan oil is high on sulfur. In general, coal from east of the Mississippi River is higher in sulfur than western coal. One method for removing SO_2 from fossil fuel combustion products involves passing the effluent gases through a device called a *scrubber*, where the gases are sprayed

The elements tellurium (left) and selenium (right).

with an aqueous suspension of calcium oxide (lime). The scrubbing eliminates most of the SO_2 but produces large amounts of $CaSO_3$ and $CaSO_4$ that must be disposed of.

8-13 SELENIUM AND TELLURIUM BEHAVE LIKE SULFUR

Selenium is found in the rare minerals *crooksite* and *clausthalite*, and tellurium occurs rarely as the free element or as the telluride of gold ($AuTe_2$) and other metals. The major commercial source of selenium and tellurium is the anode muds produced in the electrolytic refining of impure copper metal obtained from copper sulfide ores. Selenium occurs in several allotropic forms including a metallic gray hexagonal form and as deep-red, monoclinic crystals, which are composed of cyclic Se_8 molecules, analogous to S_8. Pure crystalline tellurium is very brittle and has a silvery-white metallic luster. Both selenium and tellurium are *p*-type semiconductors and are used to fabricate various photoelectric and solid-state electronic devices. Selenium is used in photocells, solar cells, and rectifiers (AC to DC current converters), and in xerography. Bismuth-tellurium semiconductors are used in thermoelectric coolers that can remove energy as heat from a liquid when an electric current is passed through the device immersed in the liquid.

The chemistries of selenium and tellurium are similar to that of sulfur in that they behave essentially like nonmetals in forming covalently bonded compounds. However, as is normally found on descending a group, there is an increase in metallic character, and tellurium has a very slight metallic character, whereas polonium is a metal. Thus, like sulfur, selenium and tellurium form hydrides and oxides analogous to H_2S, SO_2, and SO_3.

Table 8-7 Some important compounds of selenium and tellurium

Compound	*Uses*
cadmium selenide, CdSe(*s*)	produce ruby-colored glass, ceramics, and enamels
selenium dioxide, $SeO_2(s)$	antioxidant in lubricating oils
selenium sulfide, SeS(*s*)	medicated shampoos (treatment of seborrhea)
tungsten selenide, $WSe_2(s)$	as a solid lubricant for vacuum and elevated-temperature applications
cadmium telluride, CdTe(*s*)	semiconductors, phosphors, infrared-transmitting material
tellurium dioxide, $TeO_2(s)$	tinting glass

▪ Selenium sulfide is an ingredient in several anti-dandruff shampoos.

Most metallic elements react directly with Se and Te to form compounds like those obtained in the analogous reactions with sulfur. Thus we have Na_2Se, CaTe, and FeTe. The reactions of selenium and tellurium with nonmetals are also analogous to those of sulfur, as seen in the reactions

$$Se(s) + 2Cl_2(g) \rightarrow SeCl_4(l)$$

$$Te(s) + \underset{\text{excess}}{3F_2(g)} \rightarrow TeF_6(g)$$

$$Se(s) + O_2(g) \rightarrow SeO_2(g)$$

Like arsenic, selenium shows a reluctance to be oxidized to its maximum oxidation state. For example, unlike SO_3 and TeO_3, SeO_3 is thermally unstable with respect to SeO_2:

$$SeO_3(s) \rightarrow SeO_2(s) + \tfrac{1}{2}O_2(g) \qquad \Delta H^\circ_{rxn} = -58.5 \text{ kJ}$$

The acids H_2SeO_3 (selenous acid) and H_2SeO_4 (selenic acid) are prepared by dissolving SeO_2 and SeO_3, respectively, in water. These acids are analogous to sulfurous acid and sulfuric acid, respectively. However, TeO_2 is insoluble in water, and telluric acid is not at all like sulfuric acid. Its formula is $Te(OH)_6$, and it is a very weak acid ($pK_a \sim 7$ at 25°C).

Some compounds of selenium and tellurium are given in Table 8-7.

TERMS YOU SHOULD KNOW

photosynthesis
ozone layer
Frasch process
roasting of ores
quenching
contact process
oleum (fuming sulfuric acid)

QUESTIONS

8-1. What is the source of most of the oxygen in the earth's atmosphere?

8-2. Give two methods used to produce small quantities of oxygen in the laboratory.

8-3. What is the heat-producing reaction of an oxyacetylene torch?

8-4. The alkali superoxides react with CO_2 according to

$$4MO_2(s) + 2CO_2(g) \rightarrow 2M_2CO_3(s) + 3O_2(g)$$

Suggest an application of this reaction.

8-5. Outline by means of balanced chemical equations a method for the preparation of $D_2O_2(aq)$.

8-6. It has been determined that in the oxidation of H_2O_2 in aqueous solution by MnO_4^-, Ce^{4+}, and other strong oxidizing agents, the $O_2(g)$ produced comes entirely from the H_2O_2, and not from water. How could this be determined?

8-7. Describe the Frasch process.

8-8. Describe what happens at various stages when sulfur (initially in the rhombic form) is heated slowly from 90°C to 450°C.

8-9. Describe, using balanced chemical equations, the contact process for the manufacture of sulfuric acid.

8-10. Why would it be unwise to attempt to increase the acidity of the soil around plants by adding concentrated sulfuric acid?

8-11. The gas inside some tennis balls is about 50 percent SF_6 and 50 percent air. Such balls retain their bounce longer than balls charged solely with air. Suggest an explanation based on molecular size for this observation.

8-12. There are two known isomers of S_2F_2 with significantly different sulfur-sulfur bond lengths. Propose structures for the two isomers, draw the Lewis formulas.

8-13. Draw a Lewis formula for disulfuric acid, $H_2S_2O_7$ (there is an S—O—S linkage).

8-14. Draw a Lewis formula for peroxomonosulfuric acid, H_2SO_5. This acid is also known as Caro's acid.

8-15. Write the chemical equation for the analytical determination of iodine by thiosulfate.

8-16. Write a chemical equation for the tarnishing of silver.

8-17. Write the chemical equations for the reactions that account for the yellow deposits of sulfur that occur near many hot springs.

8-18. Explain why the bond angles in H_2O, H_2S, H_2Se, and H_2Te decrease from 104.5° for water to 92° for H_2S, 91° for H_2Se, and 90° for H_2Te.

8-19. The compounds SF_4 and SF_6 are both very unstable with respect to reaction with water, as shown by the following ΔG°_{rxn} values.

(a) $SF_4(g) + 2H_2O(l) \rightarrow SO_2(aq) + 4HF(aq) \qquad \Delta G^\circ_{rxn} = -282\text{ kJ}$

(b) $SF_6(g) + 4H_2O(l) \rightarrow 2H^+(aq) + SO_4^{2-}(aq) + 6HF(aq) \qquad \Delta G^\circ_{rxn} = -472\text{ kJ}$

Although $SF_4(g)$ reacts rapidly with water $SF_6(g)$ does not, being inert even to hot $NaOH(aq)$ or $HNO_3(aq)$. Consider the bonding in the two molecules, and offer an explanation for the observed difference in reactivities.

8-20. Given the thermodynamic data, $\Delta G^\circ_f[SF_4(g)] = -731.3\text{ kJ}\cdot\text{mol}^{-1}$ and $\Delta G^\circ_f[SF_6(g)] = -1105.3\text{ kJ}\cdot\text{mol}^{-1}$ at 25°C, calculate the equilibrium constant of the reaction

$$SF_4(g) + F_2(g) \rightleftharpoons SF_6(g)$$

Given that $\Delta H^\circ_{rxn} = -434.1$ kJ at 25°C, is the production of SF_6 more favored at high or low temperatures?

8-21. A 35.0-mL sample of $I_3^-(aq)$ requires 28.5 mL of 0.150 M $Na_2S_2O_3(aq)$ to react with all the $I_3^-(aq)$. Calculate the concentration of $I_3^-(aq)$ in the sample.

8-22. The concentration of ozone in oxygen-ozone mixtures can be determined by passing the gas mixture into a buffered $KI(aq)$ solution. The O_3 oxidizes $I^-(aq)$ to $I_3^-(aq)$:

$$O_3(g) + 3I^-(aq) \rightarrow O_2(g) + I_3^-(aq) + 2OH^-(aq)$$

The concentration of $I_3^-(aq)$ formed is then determined by titration with $Na_2S_2O_3(aq)$. Given that

22.50 mL of 0.0100 M $Na_2S_2O_3(aq)$ are required to titrate the $I_3^-(aq)$ in a 50.0-mL sample of KI(aq) that was equilibrated with a $O_2 + O_3$ sample, compute the moles of O_3 in the sample.

8-23. Atmospheric SO_2 can be determined by reaction with $H_2O_2(aq)$:

$$H_2O_2(aq) + SO_2(g) \rightarrow H_2SO_4(aq)$$

followed by titration of the $H_2SO_4(aq)$ produced. Given that 18.50 mL of 0.0250 M NaOH was required to neutralize the $H_2SO_4(aq)$ in a 50.0-mL $H_2O_2(aq)$ sample that was equilibrated with a sample of air containing SO_2, compute the moles of SO_2 in the air sample.

THE HALOGENS

Chlorine, bromine, and iodine.

The Group 7 elements, fluorine, chlorine, bromine, iodine, and astatine, are collectively called the *halogens*. At 25°C, fluorine is a pale yellow gas, chlorine is a green-yellow gas, bromine is a dark red liquid, and iodine is a gray-violet solid. There are no stable isotopes of astatine, all of them being radioactive. All the halogens have a pungent, irritating odor and are very poisonous. The elements exist as reactive diatomic molecules, with their reactivity decreasing with increasing atomic number.

■ Halogen means salt former.

The halogens react directly with most metals and many nonmetals. Because of the reactivity of the halogens, they do not occur as the free elements in nature; they occur primarily as

Table 9-1 Sources and uses of the Group 7 elements

Element	*Principal sources*	*Uses*
fluorine	*fluorspar,* $CaF_2(s)$	production of UF_6 for the nuclear industry, production of fluorocarbons
chlorine	*halite,* $NaCl(s)$, *sylvite,* $KCl(s)$, seawater	production of organic compounds, water purification, bleaches, flame-retardant compounds, dyes, textiles, insecticides, plastics
bromine	natural brines, salt lakes and salt beds	production of ethylene dibromide (a lead scavenger in antiknock gasoline), production of pesticides, fire-retardant materials, photography, dyestuffs
iodine	brines associated with certain oil well drillings, Chilean deposits of saltpeter, seaweeds	production of organic compounds, iodized salt and tincture of iodine

halide (F^-, Cl^-, Br^-, I^-) salts. Fluorine, the 13th most abundant element in the earth's crust, occurs primarily as the minerals *fluorite,* CaF_2, *cryolite,* Na_3AlF_6, and *fluorapatite,* $Ca_{10}F_2(PO_4)_6$. Although most fluorine in the earth's crust occurs in fluorapatite, it contains too little fluorine (3.5 percent by mass) to be a commercial source. Only fluorite is used as a source of fluorine.

Chlorine is about the twentieth most abundant element and occurs in vast evaporative deposits of *rock salt,* NaCl, and *sylvite,*

Table 9-2 Atomic properties of the Group 7 elements

Property	*Fluorine*	*Chlorine*	*Bromine*	*Iodine*
atomic number	9	17	35	53
atomic mass/amu	18.998403	35.453	79.904	126.9045
number of naturally occurring isotopes	1	2	2	1
ground-state electron configuration	$[He]2s^22p^5$	$[Ne]3s^23p^5$	$[Ar]3d^{10}4s^24p^5$	$[Kr]4d^{10}5s^25p^5$
atomic radius/pm	71	99	114	133
ionic radius/pm	136	181	195	216
bond length of X_2/pm	142	198	228	266
bond enthalpy of $X_2/kJ \cdot mol^{-1}$	155	243	192	150
ionization energy/$MJ \cdot mol^{-1}$	1.68	1.26	1.14	1.01
electronegativity	4.0	3.2	3.0	2.7

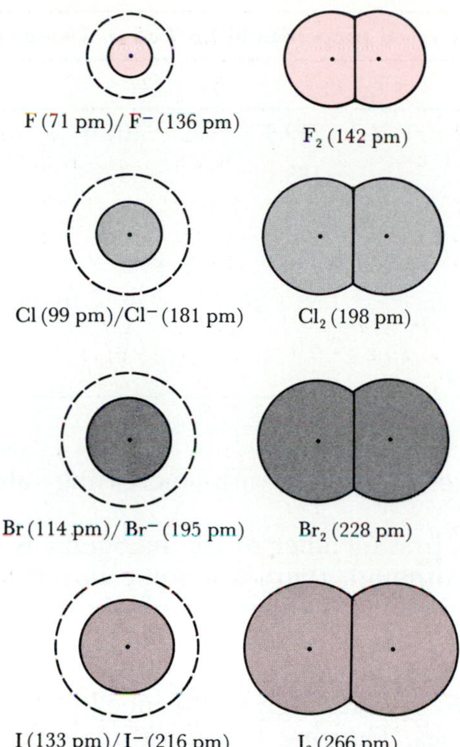

Figure 9-1 Relative sizes of neutral halogen atoms (single spheres), halide ions (dashed circles), and diatomic halogen molecules (attached pairs). Distances are in picometers.

KCl, as well as being the principal anion in seawater. The chlorine is recovered from its chloride salts by electrolysis. Bromine is only about half as abundant as chlorine. Although seawater is a potentially inexhaustible source of bromine, it is presently obtained commercially from certain natural brines and salt lakes.

Iodine is the rarest of the halogens and is about half as abundant as bromine. The principal source of iodine was once the Chilean nitrate beds, which contain a fair amount of $Ca(IO_3)_2$, but this source has been replaced by brines associated with oil wells in Louisiana and California and subterranean brines in Michigan and Oklahoma. Since World War II, however, Japan has been the world's leading producer of iodine. Table 9-1 summarizes the principal sources and commercial uses of the halogens.

The halogens serve as a good example of the variation of atomic and physical properties with atomic number. Their atomic and ionic radii and molecular bond lengths increase (Figure 9-1) and their ionization energies, electronegativities, and bond enthalpies decrease with increasing atomic number (Table 9-2). The melting points and boiling points and the values of $\Delta\overline{H}_{fus}$ and $\Delta\overline{H}_{vap}$ increase while the values of the standard re-

Table 9-3 Some physical properties of the Group 7 elements

Property	F_2	Cl_2	Br_2	I_2
melting point/°C	−219.6	−101.0	−7.2	113.5
boiling point/°C	−183.1	−34.6	58.8	184.4
$\Delta\overline{H}_{fus}$/kJ · mol^{-1}	0.510	6.41	10.6	15.5
$\Delta\overline{H}_{vap}$/kJ · mol^{-1}	6.54	20.4	29.5	41.9
density at 20°C/g · cm^{-3}	1.58×10^{-3}	2.98×10^{-3}	3.103	4.660
E^0/V at 25°C for $\frac{1}{2}X_2 + e^- \rightarrow X^-(aq)$	2.87	1.36	1.07	0.54

duction voltages decrease with increasing atomic number (Table 9-3).

Fluorine, the first member of the halogens, is the most reactive element. Although there are some important differences between the chemical properties of fluorine and the rest of the halogens, the halogens are about as uniform in their chemical properties as are the Group 1 metals. The small size of a fluorine atom and its high electronegativity account for its special properties. Fluorine is the strongest known oxidizing agent, and consequently there are no known positive oxidation states of fluorine. The only known oxidation states of fluorine are 0 and −1. For example, unlike the other halogens, fluorine forms no oxyacids. Like arsenic and selenium, bromine shows a reluctance to reach its maximum oxidation state of +7. For instance, perbromates, BrO_4^-, were not prepared until about 1970, and BrO_4^- is a stronger oxidizing agent than either ClO_4^- or IO_4^-.

9-1 FLUORINE IS THE MOST REACTIVE ELEMENT

Fluorine is a pale-yellow, corrosive gas that is the strongest oxidizing agent known and the most reactive of all the elements. It reacts directly, and in most cases vigorously, with all the elements except helium and neon. The extremely corrosive nature of fluorine is shown by its reactions with glass, ceramics, and carbon; even water burns vigorously in fluorine:

$$2F_2(g) + H_2O(g) \rightarrow OF_2(g) + 2HF(g)$$

Many of the known noble-gas compounds are fluorides, such as XeF_2, XeF_4, and XeF_6.

▪ Fluorine is the most electronegative element.

Because of its high electronegativity, fluorine is capable of stabilizing unusually high oxidation states of other elements. Some examples are

OF_2 O(II) AgF_2 Ag(II) IF_7 I(VII)

The high reactivity of fluorine is a consequence of the low F—F bond energy (139 kJ · mol^{-1}) and the high X—F bond energies (~500 kJ · mol^{-1}) to other elements. For example, the reaction

$$H_2(g) + F_2(g) \rightarrow 2HF(g) \qquad \Delta H^\circ_{rxn} = -536\ \text{kJ}$$

is highly exothermic and produces a flame temperature of over 6000°C, which is the highest known chemical flame temperature, being approximately equal to the surface temperature of the Sun.

Because of its extreme reactivity, elemental fluorine wasn't isolated until 1886. Elemental fluorine is obtained by the electrolysis of hydrogen fluoride dissolved in molten potassium fluoride:

▪ The French chemist Henri Moisson, who first isolated fluorine, received the 1906 Nobel Prize in Chemistry for his work.

$$2HF\ (in\ KF\ melt) \xrightarrow{\text{electrolysis}} H_2(g) + F_2(g)$$

The modern method of producing F_2 is essentially a variation of the method first used by Moisson. Prior to World War II, there was no commercial production of fluorine. The atomic bomb project required huge quantities of fluorine for the production of uranium hexafluoride, UF_6, a gaseous compound that is used in the separation of uranium-235 from uranium-238. It is uranium-235 that is used in nuclear devices. The production of uranium hexafluoride for the preparation of fuel for nuclear power plants is today a major commercial use of fluorine.

▪ $^{235}UF_6$ is separated from $^{238}UF_6$ by gaseous effusion. The lighter $^{235}UF_6$ effuses more rapidly than does the heavier $^{238}UF_6$.

Hydrogen fluoride is used in petroleum refining and in the production of fluorocarbon polymers, such as Teflon and Freons. It is also used to etch, or "frost," glass for light bulbs and decorative glassware via the reaction

$$SiO_2(s) + 6HF(aq) \rightarrow H_2SiF_6(s) + 2H_2O(l)$$

Because hydrofluoric acid, HF(*aq*), dissolves glass via this reaction, it must be stored in plastic bottles.

Various fluorides, such as tin(II) fluoride, SnF_2, and sodium monofluorophosphate, Na_2PO_3F, are used as toothpaste additives, and sodium fluoride is added to some municipal water supplies to aid in the prevention of tooth decay. Ordinary tooth enamel is hydroxyapatite, $Ca_{10}(OH)_2(PO_4)_6$. If low concentrations of fluoride ion are added to the diets of children, then a substantial amount of the tooth enamel formed will consist of fluorapatite, $Ca_{10}F_2(PO_4)_6$, which is much harder and less affected by acidic substances than hydroxyapatite. Consequently, fluorapatite is more resistant to tooth decay than is hydroxyapatite. The use of fluoride has decreased the incidence of tooth decay among children markedly over the past 30 years.

▪ Tin(II) fluoride is also known as stannous fluoride.

Many organofluoride compounds are used as refrigerants. Two common ones are dichlorodifluoromethane (Freon 12), CCl_2F_2, which is used in automobile air conditioners, and chlorodifluoromethane (Freon 21), $CHClF_2$, which is used in home air conditioners. Fluorocarbons have displaced refriger-

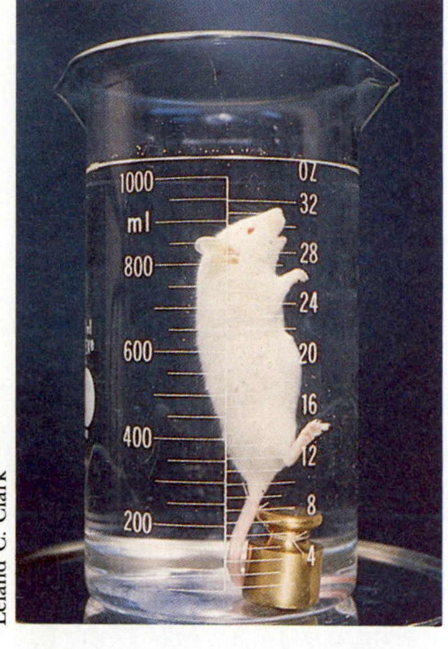

Leland C. Clark

This submerged mouse is breathing oxygen dissolved in a liquid fluorocarbon. The solubility of oxygen in this liquid is so great that the mouse is able to breath by absorbing oxygen from the oxygen-containing fluorocarbon that fills its lungs. When the mouse is removed from the liquid, the fluorocarbon vaporizes from its lungs and normal breathing resumes.

Table 9-4 Some important compounds of fluorine

Compound	*Uses*
hydrogen fluoride, HF(*g*)	catalyst in the petroleum industry; refining of uranium; etching glass; pickling stainless steel; gasoline production
boron trifluoride, $BF_3(g)$	catalyst for many organic reactions; soldering fluxes; measurement of neutron intensities
sulfur hexafluoride, $SF_6(g)$	gaseous electrical insulator in high-voltage generators and radar wave guides
sodium hexaflouroaluminate(III) (*cryolite*), $Na_3AlF_6(s)$	production of aluminum; electrical insulation; polishes; ceramics; insecticide

ants such as ammonia and sulfur dioxide in refrigerators because of their much lower toxicity and greater chemical stability. Some commercially important fluorine compounds are given in Table 9-4.

9-2 CHLORINE IS OBTAINED FROM CHLORIDES BY ELECTROLYSIS

Chlorine is a green-yellow, poisonous, corrosive gas that is prepared commercially by the electrolysis of either brines or molten rock salt:

$$\underset{\text{molten rock salt}}{2NaCl(l)} \xrightarrow{\text{electrolysis}} 2Na(l) + Cl_2(g)$$

About 10 million metric tons of chlorine are produced annually in the United States, making it the eighth ranked chemical in terms of production. It is prepared on a laboratory scale by heating a mixture of hydrochloric acid and manganese dioxide:

$$MnO_2(s) + 4H^+(aq) + 2Cl^-(aq) \rightarrow Mn^{2+}(aq) + Cl_2(g) + 2H_2O(l)$$

This reaction was used by the Swedish chemist Karl Scheele in 1774 in the first laboratory preparation of Cl_2.

Chlorine is very reactive, combining directly with most other elements with notable exceptions being carbon, nitrogen, and oxygen. Chlorine burns in hydrogen to form hydrogen chloride:

$$H_2(g) + Cl_2(g) \rightarrow 2HCl(g)$$

This reaction is used to prepare very pure HCl(*g*). Mixtures of $H_2(g)$ and $Cl_2(g)$ are explosive when exposed to light.

Many metals react directly with chlorine to form ionic chlorides:

$$Zn(s) + Cl_2(g) \rightarrow ZnCl_2(s)$$

and nonmetals react with chlorine to form covalent chlorides:

$$\underset{\text{excess}}{2Sb(s)} + 3Cl_2(g) \rightarrow 2SbCl_3(s)$$

$$2Sb(s) + \underset{\text{excess}}{5Cl_2(g)} \rightarrow 2SbCl_5(s)$$

Chlorine is a strong oxidizing agent:

$$Cl_2(g) + 2e^- \rightarrow 2Cl^-(aq) \qquad E^0 = +1.36\text{ V}$$

as illustrated by the following reactions:

$$Cl_2(g) + H_2S(aq) \rightarrow 2HCl(aq) + S(s)$$
$$Cl_2(aq) + 2Fe^{2+}(aq) \rightarrow 2Cl^-(aq) + 2Fe^{3+}(aq)$$

Most chlorine produced in the United States is used as a bleaching agent in the pulp and paper industry. It is also used extensively as a germicide in water purification and in the production of insecticides (DDT and chlordane) and herbicides (2,4-D).

Many chlorinated hydrocarbons present a serious health hazard to humans and other mammals, fishes, and birds. Such compounds are not biodegradable and, because of their high solubility in nonpolar solvents, accumulate in fatty tissues, where their presence may lead to irreversible liver damage and, in some cases, cancer. They also tend to work their way up the food chain to humans in increasing concentrations. Chlorinated hydrocarbons can be absorbed directly through the skin. An especially insidious group of chlorinated hydrocarbons are the carcinogenic PCBs, *p*oly*c*hlorinated *b*iphenyls. PCBs are inexpensive, nonflammable, very stable compounds with excellent insulation properties and, as a consequence, were once widely used in transformers and capacitors on electric power lines. The discovery of the health hazards of PCBs has led to an extensive effort to remove them from power grids. Table 9-5 lists some compounds of chlorine and their major uses.

9-3 BROMINE AND IODINE ARE OBTAINED BY OXIDATION OF BROMIDES AND IODIDES WITH CHLORINE

Bromine is a dense, red-brown, corrosive liquid with a very pungent odor. It attacks skin and tissue and produces painful, slow-healing sores. Bromine vapor and solutions of bromine in nonpolar solvents are red (Figures 9-2 and 9-3).

Table 9-5 Some important compounds of chlorine

Compound	*Uses*
hydrochloric acid, $HCl(g)$	ore refining; metallurgy; boiler-scale removal; food processing; oil and gas well treatment; general acid
sodium chloride, $NaCl(s)$	source of many sodium and chlorine compounds; food preservative; manufacture of soaps and dyes; ceramic glazes; home water softeners; highway de-icing; food seasoning; curing of hides; metallurgy
calcium chloride, $CaCl_2(s)$	drying agent; dust control on roads; paper and pulp industry; refrigeration brines; fireproofing fabrics; wood preservative; sizing and finishing cotton fabrics
potassium chlorate, $KClO_3(s)$	oxidizing agent; explosives; matches; textile printing; pyrotechnics; bleaching agent; manufacture of aniline dyes
calcium hypochlorite, $Ca(ClO)_2(s)$	algicide; bactericide; deodorant; swimming pool disinfectant; fungicide; bleaching agent for paper and textiles
sodium hypochlorite, $NaClO(s)$	bleaching agent for paper and textiles; water purification; swimming pool disinfectant; fungicide; laundry agent

The major source of bromine in the United States is from brines that contain bromide ions. The pH of the brine is adjusted to 3.5, and chlorine is added; the chlorine oxidizes bromide ion to bromine, which is swept out of the brine with a current of air:

Ethyl Corp.

Figure 9-2 Bromine processing plant. The red color is produced by the bromine gas.

Table 9-6 Some important compounds of bromine

Compound	*Uses*
sodium bromide, $NaBr(s)$	photography; sedative
potassium bromide, $KBr(s)$	gelatin bromide; photographic papers and plates; lithography; special soaps; infrared spectroscopic prisms
potassium bromate, $KBrO_3(s)$	oxidizing agent; food additive; permanent wave compound
silver bromide, $AgBr(s)$	photographic film and plates; photochromic glass
calcium bromide, $CaBr_2(s)$	photography; medicine; desiccant; food preservative; fire retardant

$$2Br^-(aq) + Cl_2(aq) \xrightarrow{pH = 3.5} 2Cl^-(aq) + Br_2(aq)$$

About 1 kg of bromine can be obtained from 15,000 L of seawater. About 200,000 metric tons of bromine were produced in the United States during 1980.

Bromine is used to prepare a wide variety of metal bromide and organobromide compounds. Its major uses are in the production of dibromoethane, $BrCH_2CH_2Br$, which is added to leaded gasolines as a lead scavenger, and in the production of silver bromide emulsions for black-and-white photographic films. Bromine is also used as a fumigant and in the synthesis of fire retardants, dyes, and pharmaceuticals, especially sedatives. Table 9-6 lists some important compounds of bromine.

Solid iodine is dark gray in color with a slight metallic luster. Iodine gas and solutions of iodine in nonpolar solvents such as carbon tetrachloride are a beautiful purple color (Figure 9-4); solutions of iodine in water and alcohols are brown as a result of the specific polar interactions between I_2 and the —O—H bond.

Iodide ion is present in seawater and is assimilated and concentrated by many marine animals and by seaweed. Certain seaweeds are an especially rich source of iodine. The iodide ion in seaweed is converted to iodine by oxidation with chlorine.

Iodine is the only halogen to occur naturally in a positive oxidation state as in the Chilean iodate deposits. The free element is obtained by reduction of IO_3^- and IO_4^- with sodium hydrogen sulfite:

$$2IO_3^-(aq) + 5HSO_3^-(aq) \rightarrow I_2(aq) + 5SO_4^{2-}(aq) + 3H^+(aq) + H_2O(l)$$

Iodine is not very soluble in pure water, but is very soluble in an aqueous KI solution. The increased solubility in KI (*aq*) is due to the formation of a linear, colorless tri-iodide species (Figure 9-4):

$$I_2(aq) + I^-(aq) \rightleftharpoons I_3^-(aq) \qquad K = 700M^{-1} \text{ at } 25°C$$

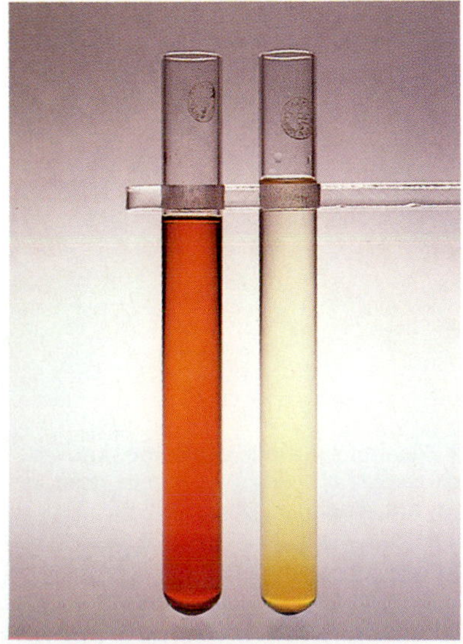

Figure 9-3 Bromine dissolved in carbon tetrachloride (left) and in water (right).

Figure 9-4 Various solutions of iodine. Left, I_2 dissolved in CCl_4. Center left, I_2 dissolved in KI(*aq*). Center right, I_2 dissolved with water. Right, I_2 dissolved in KI(aq) with starch added to the solution.

Table 9-7 Some important compounds of iodine

Compound	*Uses*
sodium iodide, NaI(*s*)	photography; feed additive; cloud seeding
potassium iodide, KI(*s*)	photographic emulsions; dietary supplement; infrared optics
silver iodide, AgI(*s*)	photography: cloud seeding
sodium iodate, $NaIO_3$(*s*)	disinfectant; antiseptic; feed additive

The presence of very low concentrations of aqueous tri-iodide can be detected by adding starch to the solution. The tri-iodide ion combines with starch to form a brilliant deep-blue species (Figure 9-4).

▪ Alcohol solutions of iodine, known as *tincture of iodine,* were once used as an antiseptic.

Iodide ion is essential for the proper functioning of the human thyroid gland, which is located in the base of the throat. Iodide deficiency manifests itself as the disease *goiter,* which causes an enlargement of the thyroid gland. Small quantities of potassium iodide are added to ordinary table salt, which is then marketed as iodized salt.

Table 9-7 lists some commercially important compounds of iodine.

9-4 HYDROGEN HALIDES ACT AS ACIDS IN AQUEOUS SOLUTION

The halogens form hydrogen halides of composition HX, where X represents a halogen. Except for HI, and to some extent for HBr, the hydrogen halides can be prepared by the reaction of concentrated sulfuric acid with the corresponding alkali metal halide, which produces gaseous HX. For example,

$$2NaCl(s) + H_2SO_4(conc) \rightarrow Na_2SO_4(s) + 2HCl(g)$$

The reaction occurs because hydrogen chloride is much less soluble in water and more volatile than sulfuric acid. A major commercial preparation of HCl is as a by-product in the chlorination of hydrocarbons, for example,

▪ About 5 billion pounds of hydrochloric acid are produced annually in the United States.

$$CH_3CH_3(g) + Cl_2(g) \rightarrow CH_3CH_2Cl(g) + HCl(g)$$

Hydrogen chloride can also be prepared on a laboratory scale by the vigorous reaction of PCl_5 with water:

$$PCl_5(s) + 4H_2O(l) \rightarrow H_3PO_4(aq) + 5HCl(g)$$

Hydrogen bromide is prepared industrially by direct reaction of hydrogen and bromine at elevated temperature in the presence of a platinum catalyst:

$$H_2(g) + Br_2(g) \xrightarrow{Pt} 2HBr(g)$$

Hydrogen iodide is oxidized to iodine by concentrated sulfuric acid and thus must be prepared by a different method. One method is the reaction of iodine with hydrazine:

$$2I_2(s) + N_2H_4(aq) \rightarrow 4HI(aq) + N_2(g)$$

Table 9-8 The hydrogen halide molar bond enthalpies

Hydrogen halide	*Molar bond enthalpy*/kJ · mol^{-1}
HF	569
HCl	431
HBr	368
HI	297

The molar bond enthalpies of the hydrogen halides decrease with increasing atomic number, as shown in Table 9-8. In water, HCl, HBr, and HI are all strong acids, and HF is a weak acid:

$$HF(aq) + H_2O(l) \rightleftharpoons H_3O^+(aq) + F^-(aq) \qquad K = 7 \times 10^{-4}\ \text{M at } 25°\text{C}$$

Hydrogen fluoride is a weak acid because of the very strong H—F bond. Because fluoride has the highest electronegativity of any element, fluorine forms the strongest hydrogen bonds. The strong hydrogen-bonding property of fluorine is seen in the formation of the linear *bifluoride* ion:

$$HF(aq) + F^-(aq) \rightleftharpoons \underset{\text{bifluoride}}{F\text{—}H\text{—}F^-(aq)} \qquad K = 5\ \text{M}^{-1} \text{ at } 25°\text{C}$$

and also by the formation of polymeric hydrogen fluorides such as $(HF)_2$ and $(HF)_6$, which are cyclic molecules.

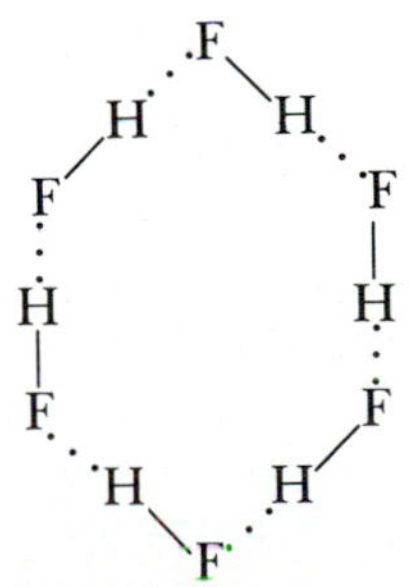

The molecule $(HF)_6$ is cyclic. Note the hydrogen bonds.

9-5 THE HALOGENS FORM NUMEROUS COMPOUNDS WITH OXYGEN-HALOGEN BONDS

The best known and most important oxygen-halogen compounds are the halogen oxyacids. The halogens form a series of oxyacids in which the oxidation state of the halogen atom can be +1, +3, +5, or +7. For example, the oxyacids of chlorine are

$HClO$	hypochlorous acid	+1
$HClO_2$	chlorous acid	+3
$HClO_3$	chloric acid	+5
$HClO_4$	perchloric acid	+7

The numbers after the names give the oxidation state of the chlorine in the acid. The Lewis formulas for these acids are

$$:\ddot{\underset{..}{Cl}}-\ddot{\underset{..}{O}}-H \qquad \ddot{Cl}(=\ddot{\underset{..}{O}})-\ddot{\underset{..}{O}}-H \qquad Cl(=\ddot{O})(=\ddot{O})-\ddot{\underset{..}{O}}-H \qquad Cl(=\ddot{O})_3-\ddot{\underset{..}{O}}-H$$

Notice that in each case the hydrogen is attached to an oxygen atom. The anions of the chlorine oxyacids are

ClO^-	hypochlorite
ClO_2^-	chlorite

Table 9-9 The halogen oxyacids and their pK_a values in water at 25°C

Halogen oxidation state	*Cl acid* pK_a	*Br acid* pK_a	*I acid* pK_a	*Acid name*	*Salt*
+1	$HClO$, 7.49	$HBrO$, 8.68	HIO, 11	hypohalous	hypohalite
+3	$HClO_2$, 1.96	—	—	halous	halite
+5	$HClO_3$, strong	$HBrO_3$, strong	HIO_3, 0.8	halic	halate
+7	$HClO_4$, strong	$HBrO_4$, strong	HIO_4, strong H_5IO_6, 3.3, 6.7	perhalic	perhalate

ClO_3^-	chlorate
ClO_4^-	perchlorate

The shapes of these ions are predicted correctly by VSEPR theory (Chapter 11 of the text) and are shown in Figure 9-5. Table 9-9 gives the known halogen oxyacids and their anions. Note that there are no oxyacids of fluorine.

When Cl_2, Br_2, or I_2 is dissolved in aqueous alkaline solution, the following type of disproportionation reaction occurs:

$$Cl_2(g) + 2OH^-(aq) \rightleftharpoons Cl^-(aq) + ClO^-(aq) + H_2O(l)$$

A solution of $NaClO(aq)$ is a bleaching agent, and many household bleaches are a 5.25% aqueous solution of sodium hypochlorite. Commercially, solutions of NaClO are manufactured by the electrolysis of cold aqueous solutions of sodium chloride:

$$2Cl^-(aq) \rightarrow Cl_2(g) + 2e^- \qquad \text{anode}$$

$$2H_2O(l) + 2e^- \rightarrow H_2(g) + 2OH^-(aq) \qquad \text{cathode}$$

The products of the two electrode reactions are allowed to mix, producing $ClO^-(aq)$ by the above reaction. Sodium hypochlorite is also employed as a disinfectant and deodorant in water supplies and sewage disposals.

Hypohalite ions decompose in basic solution via reactions of the type

$$3IO^-(aq) \xrightarrow{OH^-} 2I^-(aq) + IO_3^-(aq)$$

The analogous reaction with $ClO^-(aq)$ is slow, which makes it possible to use hypochlorite as a bleach in basic solutions. The rate of decomposition of $ClO^-(aq)$, $IO^-(aq)$, and $BrO^-(aq)$ in hot alkaline aqueous solution is sufficiently fast that when Cl_2, Br_2, or I_2 is dissolved in basic solution and the resulting solution is heated to 60°C, the following type of reaction goes essentially to completion:

$$3Br_2(aq) + 6OH^-(aq) \xrightarrow{60^\circ C} 5Br^-(aq) + BrO_3^-(aq) + 3H_2O(l)$$

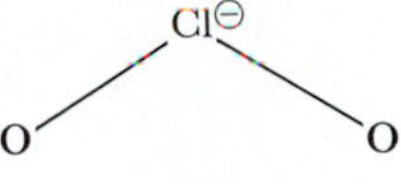

Bent

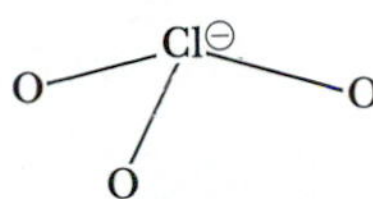

Trigonal pyramidal

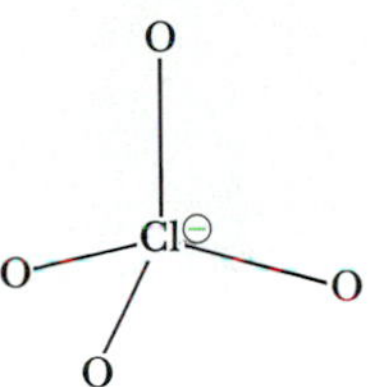

Tetrahedral

Figure 9-5 The shapes of the oxyacid anions of chlorine.

Chlorates, bromates, and iodates also can be prepared by the reaction of the appropriate halogen with concentrated nitric acid or hydrogen peroxide or (commercially) by electrolysis of the halide. For example, the reaction for the oxidation of I_2 by H_2O_2 is

$$I_2(s) + 5H_2O_2(aq) \rightarrow 2IO_3^-(aq) + 4H_2O(l) + 2H^+(aq)$$

Perchlorate and periodate are prepared by the electrochemical oxidation of chlorate and iodate, respectively. For example,

$$ClO_3^-(aq) + H_2O(l) \xrightarrow{\text{electrolysis}} ClO_4^-(aq) + 2H^+(aq) + 2e^-$$

The perchlorate is obtained from the electrolyzed cell solution by adding potassium chloride in order to precipitate potassium perchlorate, which is only moderately soluble in water. Perchloric acid, $HClO_4$, also can be obtained from the electrolyzed solution containing perchlorate by adding sulfuric acid and then distilling. The distillation is very dangerous and often results in a violent explosion. Concentrated perchloric acid should not be allowed to come into contact with reducing agents, such as organic matter, because of the extreme danger of a violent explosion. Solutions containing perchlorates should not be evaporated because of their treacherously explosive nature. Perchlorates are used in explosives, solid rocket fuels, and matches.

When potassium chlorate, $KClO_3$, is mixed with concentrated sulfuric acid and the reducing agent oxalic acid, $H_2C_2O_4$, gaseous chlorine dioxide, ClO_2, is produced:

$$2H^+(aq) + 2ClO_3^-(aq) + H_2C_2O_4(aq) \rightarrow 2ClO_2(g) + 2CO_2(g) + 2H_2O(l)$$

Chlorine dioxide is a yellow, violently explosive gas, which acts as a powerful oxidizing agent. Sodium chlorite is prepared on a commercial scale from chlorine dioxide via the reaction

$$4NaOH(aq) + Ca(OH)_2(aq) + C(s) + 4ClO_2(g) \rightarrow 4NaClO_2(aq) + CaCO_3(s) + 3H_2O(l)$$

Sodium chlorite is a strong oxidizing agent that is used in the pulp and textile industries.

9-6 INTERHALOGENS ARE COMPOUNDS CONTAINING TWO OR MORE DIFFERENT TYPES OF HALOGENS

The *interhalogen compounds* are binary compounds involving two different halogens. The known interhalogens are listed in Table 9-10. The structures of all the interhalogens are predicted correctly using VSEPR theory (Chapter 11 of the text). The interhalogens are prepared by the direct combination of the elements under appropriate experimental conditions. For example, ClF may be prepared by direct combination of the elements at a temperature of 220° to 250°C. Chlorine trifluoride, ClF_3, is also formed in the reaction, and the ClF and ClF_3 may be separated by distillation.

Most of the diatomic interhalogens disproportionate. For example, BrF disproportionates according to

$$3BrF(g) \rightleftharpoons BrF_3(g) + Br_2(l)$$

The interhalogens are all rather reactive and are strong oxidizing agents. They react with most elements to produce a mixture of halides. They react readily with water, and in some cases (for example, BrF_5) explosively.

Several polyatomic halide and interhalogen ions are known. The linear, triatomic halogen anions such as $I_3^-(aq)$ and $ICl_2^-(aq)$ are formed in reactions of the type

$$ICl(aq) + Cl^-(aq) \rightleftharpoons ICl_2^-(aq) \qquad K = 1.7 \times 10^2\ M^{-1} \text{ at } 25°C$$

Interhalogen polyhalide ions such as ICl_4^- (square planar) and BrF_6^- (octahedral) are prepared from the neutral interhalogens using reactions of the type

$$CsCl(s) + ICl_3(g) \rightarrow CsICl_4(s)$$
$$CsF(s) + BF_5(g) \rightarrow CsBrF_6(s)$$

Table 9-10 Some of the known interhalogen compounds

ClF	ClF_3	ClF_5	
BrF	BrF_3	BrF_5	
IF	IF_3	IF_5	IF_7
BrCl			
ICl			
IBr			

The structures of the polyhalide ions are predicted correctly from VSEPR theory (Figure 9-6).

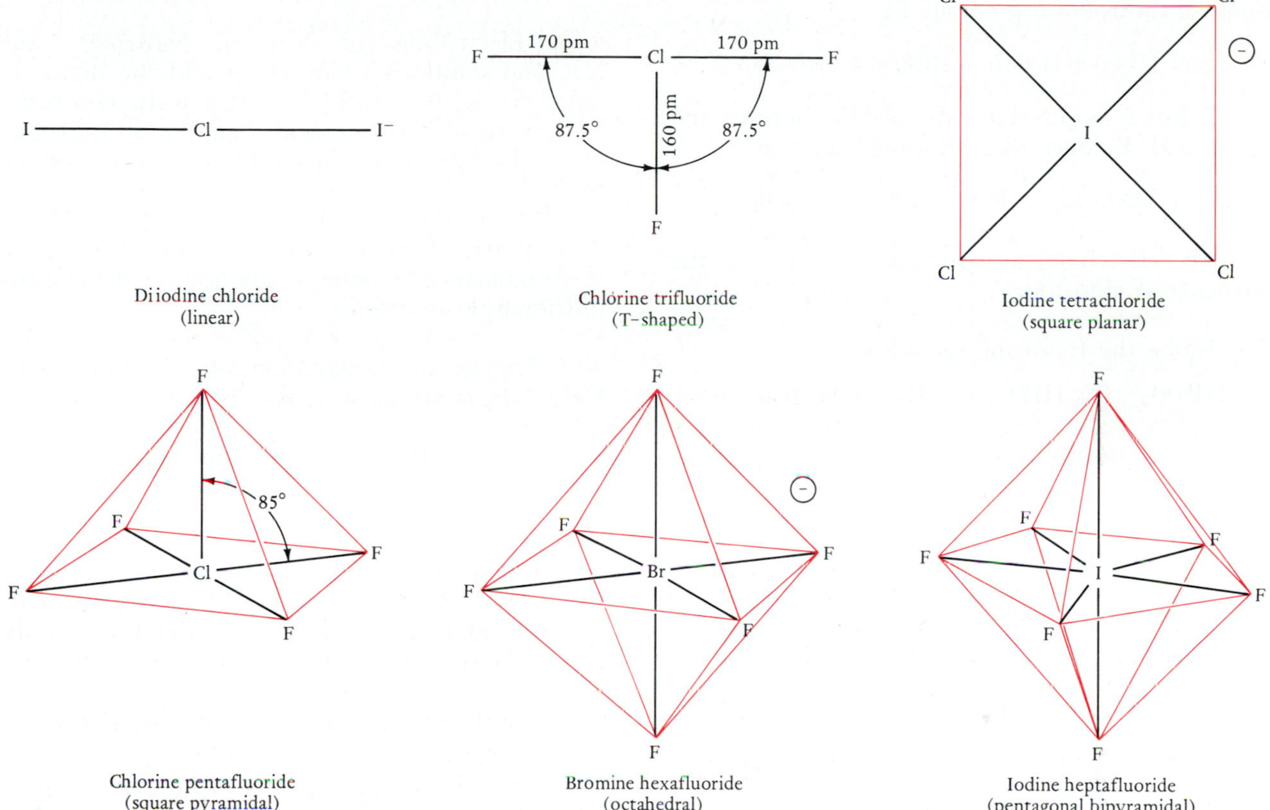

Figure 9-6 Structures of representative interhalogen species.

TERMS YOU SHOULD KNOW

halogen
tincture of iodine
hydrohalic acid
bifluoride ion
disproportionation reaction
hypohalous acid
interhalogen compound

QUESTIONS

9-1. Explain briefly why fluorine is capable of stabilizing unusually high oxidation states in many elements.

9-2. Describe how each of the halogens is prepared on a commercial scale.

9-3. Complete and balance the following equations.

(a) $Ca(s) + Br_2(l) \rightarrow$

(b) $Ti(s) + Cl_2(g) \rightarrow$

(c) $As(s) + \underset{\text{excess}}{Cl_2(g)} \rightarrow$

(d) $Na(s) + I_2(s) \rightarrow$

9-4. Complete and balance the following equations.

(a) $F_2(g) + H_2O(l) \rightarrow$

(b) $F_2(g) + Si(s) \rightarrow$

(c) $F_2(g) + Ag(s) \rightarrow$

(d) $F_2(g) + Mg(s) \rightarrow$

9-5. Balance the following equations.

(a) $NaCl(aq) + H_2SO_4(aq) + MnO_2(s) \rightarrow Na_2SO_4(aq) + MnCl_2(aq) + H_2O(l) + Cl_2(g)$

(b) $NaIO_3(aq) + NaHSO_3(aq) \rightarrow I_2(s) + Na_2SO_4(aq) + H_2SO_4(aq) + H_2O(l)$

(c) $Br_2(l) + NaOH(aq) \rightarrow$
$NaBr(aq) + NaBrO_3(aq) + H_2O(l)$

9-6. Chlorine oxidizes iodine to iodic acid in water. Balance the following equation:

$$Cl_2(g) + I_2(s) + H_2O(l) \rightarrow HCl(aq) + HIO_3(aq)$$

9-7. Iodine is oxidized to iodic acid by concentrated nitric acid. Balance the following equation:

$$I_2(s) + HNO_3(aq) \rightarrow HIO_3 + NO(g) + H_2O(l)$$

9-8. Give the chemical formulas and names of the oxyacids of chlorine.

9-9. Name the following oxyacids.

(a) $HBrO_2$ (b) HIO (c) $HBrO_4$ (d) HIO_3

9-10. Name the following oxyacids:

(a) HNO_2 (c) H_3PO_2 (e) $H_2N_2O_2$

(b) H_2SO_3 (d) H_3PO_3

and the following salts:

(a) K_2SO_3 (c) KIO_2

(b) $Ca(NO_2)_2$ (d) $Mg(BrO)_2$

9-11. Describe by balanced chemical equations how you would prepare $KIO_3(s)$ starting with $I_2(s)$.

9-12. Why is the heat evolved per mole in the neutralization reaction of $HCl(aq)$ by $KOH(aq)$ the same as that for $HBr(aq)$ by $KOH(aq)$? Why is the heat evolved per mole much less for the reaction $HF(aq)$ plus $KOH(aq)$?

9-13. In addition to the (meta)periodate ion, IO_4^-, there are the ions $H_2IO_5^-(aq)$ and $H_4IO_6^-(aq)$ that can be viewed as mono- and dihydrates, respectively, of IO_4^-. These species are formed in acidic solutions. In strongly acidic solutions, the principal I(VII) species is paraperiodic acid, $H_5IO_6(aq)$.

(a) Write a balanced chemical equation for the formation of $H_5IO_6(aq)$ from $IO_4^-(aq)$.

(b) Paraperiodic acid is a powerful oxidizing agent that can oxidize $Mn^{2+}(aq)$ to $MnO_4^-(aq)$. Write a balanced chemical equation for this reaction. Assume that the iodine product is $IO_3^-(aq)$.

9-14. The rate of disproportionation of $I_2(aq)$ is fast at all temperatures, and the following reaction occurs rapidly and quantitatively:

$I_2(aq) + OH^-(aq) \rightarrow$
$I^-(aq) + IO_3^-(aq) + H_2O(l)$ (unbalanced)

Balance this equation.

9-15. Electrolysis of $NaI(aq)$, $NaBr(aq)$, and $NaCl(aq)$ solutions yields $H_2(g)$ and the diatomic halogens, I_2, Br_2, and Cl_2, respectively. However, electrolysis of $NaF(aq)$ yields $H_2(g)$ and $O_2(g)$. Explain why $F_2(g)$ is not formed in the electrolysis of $NaF(aq)$.

9-16. When $IF(g)$ is heated it disproportionates. Write a balanced chemical equation for the disproportionation reaction.

9-17. Suggest an explanation why, in contrast to NH_3, NF_3 is not at all basic.

9-18. What is the oxidation state of the oxygen atom in HOF, an unstable substance that decomposes to HF and O_2?

9-19. When perchloric acid is dehydrated by P_4O_{10}, a colorless, unstable oily liquid, Cl_2O_7, is produced. Use hybrid orbitals to describe the bonding in Cl_2O_7 (there is a Cl—O—Cl bond).

9-20. Use VSEPR theory to predict the shapes of the following interhalogen cations.

(a) ClF_2^+ (b) ClF_4^+ (c) ClF_6^+

9-21. Name the oxidation state of each halogen in the following compounds.

(a) IF_5 (d) ClF

(b) NaClO (e) $NaIO_3$

(c) $KBrO_3$

9-22. The acid $HF(aq)$ differs from the other hydrohalic acids in that it is a weak acid (25°C data):

$$HF(aq) + H_2O(l) \rightleftharpoons H_3O^+(aq) + F^-(aq) \qquad pK_a = 3.17$$

and in that the ion $HF_2^-(aq)$ forms readily:

$$HF(aq) + F^-(aq) \rightleftharpoons HF_2^-(aq) \qquad K = 5.1\ M^{-1}$$

Suppose we have a solution with a stoichiometric concentration of $HF(aq)$ of 0.10 M that is buffered at pH = 3.00. Compute the concentrations of $F^-(aq)$, $HF(aq)$, and $HF_2^-(aq)$ in the solution.

9-23. Given that $\Delta \overline{G}_f^\circ[I_2(aq)] = 16.40\ kJ \cdot mol^{-1}$, $\Delta \overline{G}_f^\circ[I^-(aq)] = -51.57\ kJ \cdot mol^{-1}$, and $\Delta \overline{G}_f^\circ[I_3^-(aq)] =$

$-51.40\ \text{kJ} \cdot \text{mol}^{-1}$ at 25°C, calculate the equilibrium constant for the reaction

$$I_2(aq) + I^-(aq) \rightleftharpoons I_3^-(aq)$$

9-24. Given that $\Delta\overline{G}_f^\circ[\text{ICl}(aq)] = -17.1\ \text{kJ} \cdot \text{mol}^{-1}$, $\Delta\overline{G}_f^\circ[\text{Cl}^-(aq)] = -131.23\ \text{kJ} \cdot \text{mol}^{-1}$, and $\Delta\overline{G}_f^\circ[\text{ICl}_2^-(aq)] = -161.0\ \text{kJ} \cdot \text{mol}^{-1}$ at 25°C, calculate the equilibrium constant for the reaction

$$\text{ICl}(aq) + \text{Cl}^-(aq) \rightleftharpoons \text{ICl}_2^-(aq)$$

9-25. Iodine pentoxide is a reagent for the quantitative determination of carbon monoxide. The reaction is

$$5\text{CO}(g) + I_2O_5(s) \rightarrow I_2(s) + 5\text{CO}_2(g)$$

The iodine produced is dissolved in KI(aq) and then determined by reaction with $Na_2S_2O_3$:

$$2S_2O_3^{2-}(aq) + I_3^-(aq) \rightarrow 3I^-(aq) + S_4O_6^{2-}(aq)$$

Compute the moles of CO required to produce sufficient $I_3^-(aq)$ to react completely with the $S_2O_3^{2-}(aq)$ in 10.0 mL of 0.0350 M $Na_2S_2O_3(aq)$.

9-26. The solubility of $I_2(s)$ in water at 25°C is 0.0013 M and

$$I_2(aq) + I^-(aq) \rightleftharpoons I_3^-(aq) \qquad K = 700\ \text{M}^{-1}$$

at 25°C. Compute the solubility of $I_2(s)$ in a solution that is initially 0.10 M in KI(aq).

9-27. Using the standard reduction potentials

$$H^+(aq) + \text{HOCl}(aq) + e^- \rightleftharpoons \tfrac{1}{2}\text{Cl}_2(g) + H_2O(l) \qquad E^0 = 1.63\ \text{V}$$

$$\tfrac{1}{2}\text{Cl}_2(g) + e^- \rightleftharpoons \text{Cl}^-(aq) \qquad E^0 = 1.36\ \text{V}$$

calculate the value of the equilibrium constant at 25°C for the reaction

$$\text{Cl}_2(g) + H_2O(l) \rightleftharpoons H^+(aq) + \text{Cl}^-(aq) + \text{HOCl}(aq)$$

9-28. Using the result of Question 9-27, and given that

$$\text{Cl}_2(g) \rightleftharpoons \text{Cl}_2(aq) \qquad K = 0.062\ \text{M} \cdot \text{atm}^{-1}$$

calculate the value of the equilibrium constant for

$$\text{Cl}_2(aq) + H_2O(l) \rightleftharpoons H^+(aq) + \text{Cl}^-(aq) + \text{HOCl}(aq)$$

THE NOBLE GASES

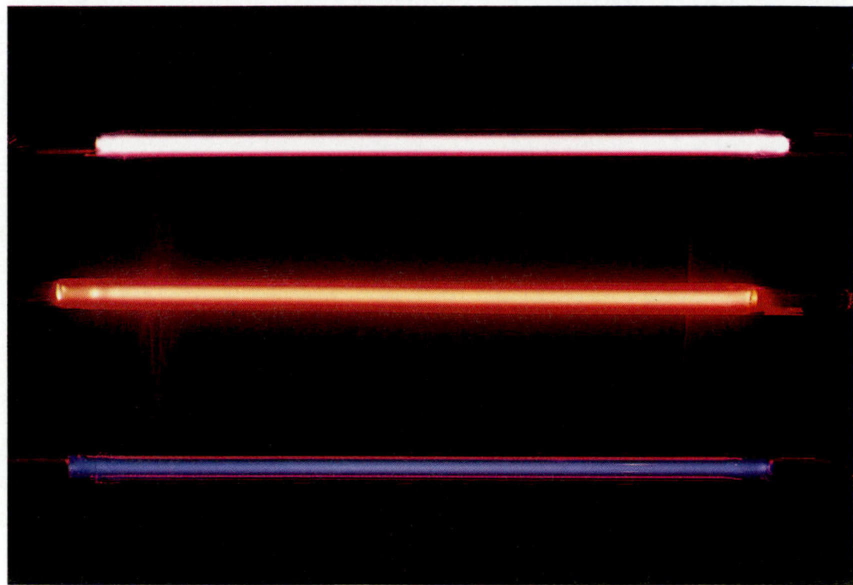

When an electric discharge is passed through a noble gas, light is emitted as electronically excited noble-gas atoms decay to lower energy levels. The tubes contain helium (top), neon (center), and argon (bottom).

The Group 8 elements, helium, neon, argon, krypton, and xenon, are called the noble gases and are noteworthy for their relative lack of chemical reactivity. Only xenon and krypton are known to enter into chemical compounds, and even then only with the two most electronegative compounds, fluorine and oxygen. As Table 10-1 indicates, the principal source of the noble gases, except for helium, is the atmosphere. The atomic and physical properties of the noble gases are given in Tables 10-2 and 10-3, respectively.

The data in Table 10-3 nicely illustrate trends in physical properties with increasing atomic size. Note that the boiling points

Table 10-1 Sources and uses of the noble gases.

Element	*Principal sources*	*Uses*
helium	natural gas wells	provide an inert atmosphere for welding, inflation of meteorlogical balloons, cryogenic, coolant, nitrogen substitute for SCBA (self-contained breathing apparatus), pressurize liquid fuel rockets
neon	fractional distillation of liquid air	fluorescent tubes, lasers, high-voltage indicators, cryogenic research
argon	fractional distillation of liquid air	provide an inert atmosphere for welding, fluorescent tubes, blanketing material for the production of titanium and other metals, lasers, deaeration of solutions
krypton	fractional distillation of liquid air	fluorescent tubes, high-speed photographic lamps, lasers
xenon	fractional distillation of liquid air	fluorescent tubes, lasers, stroboscopic lamps

and the enthalpies of vaporization increase with increasing atomic number, due to the larger London attractive forces (see Section 13-4 of the text). This same tendency can be seen in the increase of the van der Waals constants for helium through xenon (see Table 5-5 of the text). Recall that the van der Waals constants a and b are a measure of the attraction and the size, respectively, of the molecules in a gas.

10-1 THE NOBLE GASES WERE NOT DISCOVERED UNTIL 1893

In 1893, the English physicist Lord Rayleigh noticed a small discrepancy between the density of nitrogen obtained by the removal of oxygen, water vapor, and carbon dioxide from air and

Table 10-2 Atomic properties of the Group 8 elements

Property	*Helium*	*Neon*	*Argon*	*Krypton*	*Xenon*	*Radon*
atomic number	2	10	18	36	54	86
atomic mass/amu	4.00260	20.179	39.948	83.80	131.30	(222)
number of naturally occurring isotopes	2	3	3	6	9	0
outer shell electron configuration	$1s^2$	$2s^22p^6$	$3s^23p^6$	$4s^24p^6$	$5s^25p^6$	$6s^26p^6$
atomic radius/pm	50	70	95	110	130	—
ionization energy/$MJ \cdot mol^{-1}$	2.37	2.08	1.52	1.35	1.17	1.04

Table 10-3 Some physical properties of the Group 8 elements

property	*Helium*	*Neon*	*Argon*	*Krypton*	*Xenon*
melting point/°C	—	−248.6	−189.4	−157.2	−111.8
melting point/K	—	24.6	83.8	115.9	161.3
boiling point/°C	−268.9	−246.1	−185.9	−153.4	−108.1
boiling point/K	4.2	27.1	87.3	119.7	165.0
$\Delta\overline{H}_{fus}$/kJ · mol^{-1}	—	0.335	1.17	1.63	2.30
$\Delta\overline{H}_{vap}$/kJ · mol^{-1}	0.08	1.76	6.52	9.03	12.63
density at 20°C/g · cm^{-3}	1.66×10^{-4}	8.37×10^{-4}	1.66×10^{-3}	3.48×10^{-3}	5.46×10^{-3}
ppm in air	5.2	18.2	9340	1.1	0.08

the density of nitrogen prepared by chemical reaction, such as the thermal decomposition of ammonium nitrite:

$$NH_4NO_2(s) \rightarrow N_2(g) + 2H_2O(g)$$

One liter of nitrogen at 0°C and 1 atm obtained by the removal of all the other known gases from air (Figure 10-1) has a mass of 1.2572 g, whereas one liter of dry nitrogen obtained from ammonium nitrite has a mass of 1.2505 g under the same conditions. This slight difference led Lord Rayleigh to suspect that some other gas was present in the sample of nitrogen obtained from air.

The English chemist William Ramsay found that if hot calcium metal is placed in a sample of nitrogen obtained from air, about 1 percent of the gas fails to react. Pure nitrogen would

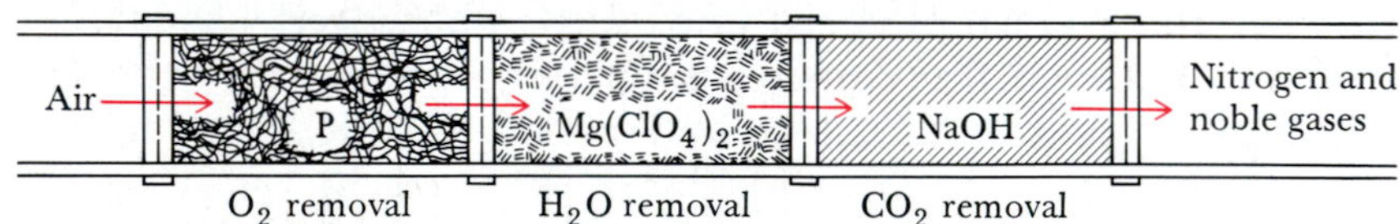

Figure 10-1 A schematic illustration of the removal of O_2, H_2O, and CO_2 from air. First the oxygen is removed by allowing the air to pass over phosphorus:

$$P_4(s) + 5O_2(g) \rightarrow P_4O_{10}(s)$$

The residual air is passed through anhydrous magnesium perchlorate to remove the water vapor:

$$Mg(ClO_4)_2(s) + 6H_2O(g) \rightarrow Mg(ClO_4)_2 \cdot 6H_2O(s)$$

and then through sodium hydroxide to remove the CO_2:

$$NaOH(s) + CO_2(g) \rightarrow NaHCO_3(s)$$

The gas that remains is primarily nitrogen with about 1 percent noble gases.

react completely. Because of the inertness of the residual gas, Ramsay gave it the name argon (Greek, idle). Ramsay then liquefied the residual gas and, upon measuring its boiling point, discovered that it consisted of five components, each with its own characteristic boiling point (Table 10-3). The component present in the greatest amount retained the name argon. The others were named helium (sun), neon (new), krypton (hidden), and xenon (stranger). Helium was named after the Greek word for sun (helios) because its presence in the sun had been determined earlier by spectroscopic methods.

The noble gases in the atmosphere are thought to have arisen as by-products of the decay of radioactive elements in the earth's crust (Chapter 24). For their work in discovering and characterizing an entire new family of elements, Rayleigh received the 1904 Nobel Price in physics and Ramsay received the 1904 Nobel Prize in chemistry.

All the noble gases are colorless, odorless, and relatively inert. Helium is used in lighter-than-air craft, despite the fact that it is denser and hence has less lifting power than hydrogen, because it is nonflammable. Helium is also used in welding to provide an inert atmosphere around the welding flame and thus reduce corrosion of the heated metal. Neon is used in neon signs, which are essentially discharge tubes (see Frontispiece) filled with a noble gas or a noble-gas mixture. When placed in a discharge tube, neon emits an orange-yellow glow that penetrates fog very well. Argon, the most plentiful and least expensive noble gas, often is used in fluorescent and incandescent light bulbs because it does not react with the discharge electrodes or the hot filament. Krypton and xenon are scarce and costly, which limits their application, although they are used in lasers, flashtubes for high-speed photography, and automobile-engine timing lights.

▪ Helium has 93 percent of the lifting power of hydrogen.

10-2 XENON FORMS COMPOUNDS WITH FLUORINE AND OXYGEN

Prior to 1962 most chemists believed, and essentially all general chemistry textbooks proclaimed, that the noble gases did not form any chemical compounds. In fact, the gases helium through xenon were called the inert gases, indicating that they did not undergo any chemical reactions.

In 1962, Neil Bartlett, then of the University of British Columbia, was working with the extremely strong oxidizing agent platinum hexafluoride, PtF_6, which oxidizes $O_2(g)$ to produce the ionic compound $O_2^+PtF_6^-$:

$$O_2(g) + PtF_6(s) \rightarrow O_2^+PtF_6^-$$

Bartlett observed that the ionization energy of $O_2(g)$ (1171 kJ · mol^{-1}) is about the same as the ionization energy of Xe(g) (1176 kJ · mol^{-1}), and so reasoned that xenon might react with PtF_6

▪ Linus Pauling predicted xenon compounds in the 1930's on the basis of the similarity of the ionization energies of O_2 and Xe.

in an analogous manner. When he mixed xenon and PtF_6 in a reaction chamber, he obtained a definite chemical reaction that at the time was thought to be the formation of $Xe^+PtF_6^-$. It has since been found that the product of the reaction is more complex than $Xe^+PtF_6^-$, but nevertheless Bartlett showed that xenon will react with a strong oxidizing agent under the right conditions of temperature and pressure. Bartlett's discovery prompted other research groups to investigate reactions of xenon, and within a year or so several other compounds of xenon were synthesized. Three xenon fluorides can be prepared by the direct combination of xenon and fluorine in a nickel vessel:

$$Xe(g) + F_2(g) \rightleftharpoons XeF_2(s)$$

$$XeF_2(s) + F_2(g) \rightleftharpoons XeF_4(s)$$

$$XeF_4(s) + F_2(g) \rightleftharpoons XeF_6(s)$$

As these three equilibria indicate, the reaction of a mixture of xenon and fluorine yields a mixture of XeF_2, XeF_4, and XeF_6. The chief difficulty is the separation of the products. A favorable yield of XeF_2 can be obtained by using a large excess of xenon. Xenon difluoride forms large, colorless crystals that melt at 130°C. It is a linear molecule, as predicted by VSEPR theory (AX_2E_3). Xenon difluoride is soluble in water and evidently exists as XeF_2 molecules in solution. Xenon tetrafluoride can be obtained in quantitative yield by reacting a 1:5 mixture of Xe and F_2 at 400°C and 6 atm in a nickel vessel. Xenon tetrafluoride forms colorless crystals (see Figure 10-2) that melt at 177°C. The molecule is square planar, as predicted by VSEPR theory (AX_4E_2). Unlike XeF_2, XeF_4 hydrolyzes according to

$$6XeF_4(s) + 12H_2O(l) \rightleftharpoons 2XeO_3(s) + 4Xe(g) + 3O_2(g) + 24HF(aq)$$

An aqueous solution of XeO_3 is stable, colorless, and odorless, and a powerful oxidizing agent. Upon evaporation to dryness $XeO_3(s)$ results. Xenon trioxide is extremely explosive, which is why work with XeF_4 (and also XeF_6) must be done under carefully dry conditions.

Argonne Nat. Lab.

Figure 10-2 Xenon tetrafluoride crystals. Xenon tetrafluoride was first prepared in 1962 by the direct combination of $Xe(g)$ and $F_2(g)$ at 6 atm and 400°C.

Some other known compounds of xenon are given in Table 10-4. Note that xenon forms chemical bonds with the most electronegative elements, fluorine and oxygen, and exhibits oxidation states of +2, +4, +6, and +8. Xenon, having the greatest atomic size of any of the nonradioactive noble gases, has the smallest ionization energy. Hence, except for radon, xenon is the most "reactive" noble gas, and we except the reactivity of the noble gases to decrease from xenon to helium. The only known molecule containing krypton is KrF_2, and no isolable compounds of argon have been reported. The molar enthalpies of formation of XeF_2 and KrF_2 are

$$Xe(g) + F_2(g) \rightleftharpoons XeF_2(s) \qquad \Delta\overline{H}_f^\circ \simeq -110\ \mathrm{kJ\cdot mol^{-1}}$$

$$Kr(g) + F_2(g) \rightleftharpoons KrF_2(s) \qquad \Delta\overline{H}_f^\circ \simeq +60\ \mathrm{kJ\cdot mol^{-1}}$$

Table 10-4 The principal compounds of xenon

Compound	*Oxidation state*	*Physical state*	*Molecular shape*
XeF_2	+2	colorless crystals	linear
XeF_4	+4	colorless crystals	square planar
XeF_6	+6	colorless crystals	distorted octahedron
$XeOF_4$	+6	colorless liquid	square pyramidal
XeO_2F_2	+6	colorless crystals	seesaw
XeO_3	+6	colorless crystals	trigonal pyramidal
XeO_4	+8	colorless gas	tetrahedral

The difference in the values of $\Delta \overline{H}_f^\circ$ for these two reactions can be accounted for by the difference in the ionization energies of krypton and xenon (180 $kJ \cdot mol^{-1}$). Although radon has the lowest ionization energy of all the noble gases and might be expected to be the most reactive, its radioactivity makes the study of radon chemistry difficult, so little is known.

QUESTIONS

10-1. What is the principal chemical property of the noble gases?

10-2. Discuss how the noble gases were discovered by Lord Rayleigh.

10-3. What is the principal source of helium?

10-4. Why did Rayleigh and Ramsey place the newly discovered noble gases in a new group in the periodic table?

10-5. What is the source of He(*g*) in natural gas deposits?

10-6. Sketch an experimental set up for removing O_2, H_2O, and CO_2 from air. How could you remove the remaining N_2?

10-7. Describe the important role that the noble gases played in the theories of chemical bonding and electronic structure of atoms.

10-8. Nitrogen is also a relatively inert gas. Suggest an experiment to demonstrate the difference between nitrogen and argon.

10-9. When Bartlett prepared $O_2^+PtF_6^-$ in 1962, what reasoning did he use to conjecture that it might be possible to prepare $Xe^+PtF_6^-$?

10-10. Complete and balance.

(a) $XeF_4(s) + H_2O(l) \rightarrow$

(b) $Xe(g) + F_2(g) \rightarrow$

(c) $Kr(g) + F_2(g) \rightarrow$

(d) $XeF_6(s) + H_2O(l) \rightarrow$

10-11. Use the data in Table 10-3 to compute the values of $\Delta \overline{S}_{vap}$ and $\Delta \overline{S}_{fus}$ for the Group 8 elements. Compare your results with the values of $\Delta \overline{S}_{vap}$ from Trouton's rule.

10-12. Why do both van der Waals constants, *a* and *b*, increase with increasing atomic number for the noble gases?

10-13. Why does $\Delta \overline{H}_{vap}$ increase with increasing atomic number for the noble gases?

10-14. Use VSEPR theory to predict the shapes of the xenon compounds given in Table 10-4.

10-15. What is the oxidation state of xenon in each of the compounds in Table 10-4?

10-16. Use VSEPR theory to predict the structures of the following compounds.
(a) RnF_2 (b) RnF_4 (c) RnO_3 (d) RnO_4

THE TRANSITION METALS

The 3*d* transition series metals. Top row from left to right: Ti, Zn, Cu, Ni, and Co. Bottom row: Sc, V, Cr, Mn, and Fe. The elements are not arranged in order of the periodic table.

In this chapter we describe the sources, chemical properties, and uses of the transition metals. Transition metal alloys are the structural backbone of modern civilization. Human development progressed from the Stone Age to the Bronze Age and then to the Iron Age. The Industrial Revolution was powered by steam engines made from steels. The Space and Computer Age utilizes a remarkable variety of exotic alloys developed to meet a wide range of specialized requirements.

The transition metals that we shall discuss in this chapter are the *d* transition metals, or the *d*-block elements. They are called the 3*d* transition-metal series, the 4*d* transition-metal series, the

5*d* series, and so forth, to indicate the subshell that is being filled within the series. There are ten members of each *d* transition-metal series, corresponding to the ability of a *d* subshell to hold a maximum of ten electrons. The ground-state outer electron configurations of the 3*d*, 4*d*, and 5*d* transition-metal series are given in Table 8-6 of the text. Note that the filling of the *d* orbitals is not perfectly regular in all cases. For example, the ground-state electron configuration of chromium is $3d^5 4s^1$ rather than $3d^4 4s^2$. These occasional irregularities arise because the *nd* and $(n + 1)s$ (for example, 3*d* and 4*s*) orbitals for a given series have energies that are fairly close together, and that interchange order as we move across a series.

Many of the transition metals are probably familiar to you. Iron, nickel, chromium, tungsten, and titanium are widely used in alloys for structural materials and play a key role in the world's technology. The precious metals, gold, platinum, and silver, are used as hard currency, in dentistry, and to make jewelry and electrical components. Copper is the most widely used metal for electric wiring. Mercury is the only metallic element that is a liquid at room temperature.

The transition metals vary greatly in abundance. Iron and titanium are the fourth and tenth most abundant elements in the earth's crust, whereas rhenium (Re) and hafnium (Hf) are very rare. The characteristics of the transition metals vary from family to family; however, they are characterized by high densities and high melting points. The two metals with the greatest densities (iridium, Ir, $22.65\ g \cdot cm^{-3}$, and osmium, Os, $22.61\ g \cdot cm^{-3}$) and the metal with the highest melting point (tungsten, W, 3410°C) are transition metals.

The physical properties, principal sources and major uses of the *d* transition series metals are given in Tables 11-1 through 11-3.

We shall not discuss the chemistry of all the transition metals, but we shall focus on the first transition-metal series. Although some comparisons with the other transition-metal series can be made, like the main-group families, where the first member differs significantly from the others, the chemistry of the first transition-metal series differs appreciably from that of the others. In particular, the aqueous solution chemistry of the first transition-metal series is simpler than that of the heavier transition metals.

11-1 THE MAXIMUM OXIDATION STATES OF SCANDIUM THROUGH MANGANESE ARE EQUAL TO THE TOTAL NUMBER OF 4*s* AND 3*d* ELECTRONS

The chemistry of even the first transition metal series is especially rich due to the several oxidation states available to many of the metals (Table 11-4). In spite of the differences in the chemistries of the transition metals, certain trends do exist:

Table 11-1 Properties of the 3*d* transition series metals

Element	*Density*/g · cm^{-3}	*Melting point*/°C	*Principal sources*	*Uses*
Sc	3.0	1541	thortveitite, $(Sc,Y)_2Si_2O_7$	no major industrial uses
Ti	4.5	1660	rutile, TiO_2	high-temperature, lightweight steel alloys, TiO_2 in white paints
V	6.0	1890	vanadinite, $(PbO)_9(V_2O_5)_3PbCl_2$	vanadium steels (rust-resistant)
Cr	7.2	1860	chromite, $FeCr_2O_4$	stainless steels, chrome plating
Mn	7.4	1244	pyrolusite, MnO_2 manganosite, MnO nodules on ocean floor	alloys
Fe	7.9	1535	hematite, Fe_2O_3 magnetite, Fe_3O_4	steels
Co	8.9	1490	cobaltite, $CoS_2 \cdot CoAs_2$ linnaetite, Co_3S_4	alloys, cobalt-60 medicine
Ni	8.9	1455	pentlandite, $(Fe, Ni)_9S_8$ pyrrhotite, $F_{0.8}S$	nickel plating, coins, magnets, catalysts
Cu	9.0	1083	chalcopyrite, $CuFeS_2$ cuprite, Cu_2O malachite, $Cu_2(CO_3)(OH)_2$	bronzes, brass, coins, electric conductors
Zn	7.1	420	zinc blende, ZnS smithsonite, $ZnCO_3$	galvanizing, bronze, brass, dry cells

1. For scandium through manganese, the highest oxidation state is equal to the total number of 4*s* plus 3*d* electrons, and this oxidation state is achieved primarily in oxygen compounds or in fluorides and chlorides. Furthermore, the stability of the highest oxidation state decreases from scandium to manganese. Thus we have

Sc_2O_3	a stable oxide
TiO_2	a stable, common ore of titanium
V_2O_5	mild oxidizing agent
CrO_4^{2-}	strong oxidizing agent
MnO_4^-	strong oxidizing agent

2. Except for scandium and titanium, all the 3*d* transition metals form divalent ions in aqueous solution.
3. For a given metal, the oxides become more acidic with increasing oxidation state.
4. For a given metal, the halides become more covalent with increasing oxidation state and hydrolyze more readily.

Let's briefly discuss some of the chemical properties of the first transition metal series.

Scandium

The ground-state electron configuration of scandium is $[Ar]3d^14s^2$, and it is somewhat similar to aluminum in its

Table 11-2 Properties of the 4*d* transition series metals

Element	*Density*/g · gm^{-3}	*Melting point*/°C	*Principal sources*	*Uses*
Y	4.5	1522	rare-earth ores	coating on high-temperature alloys, microwave ferrites, special semiconductors
Zr	6.5	1852	monazite (with rare earths) zircons, $ZrSO_4$	fuel-rod cladding for nuclear reactors, explosive primers, formerly in antiperspirants
Nb	8.6	2468	columbite, $(Fe,Mn)/(Nb,Ta)_2O_6$	stainless steels, welding rods, nuclear reactor alloys, superconductors
Mo	10.3	2622	molybdenite, MoS_2 wulfenite, $PbMoO_4$	tool steels, boiler plate, rifle barrels, spark plugs, X-ray tube filaments
Tc	11.5	2172	does not occur in nature	brain and thyroid scans (Tc-99)
Ru	12.4	2310	osmiridium platinum ores	platinum substitute in jewelry, pen nibs, electrical contact alloys, catalyst, superconductors
Rh	12.4	1966	platinum ores rhodite	platinum alloys, plating for kitchen utensils, reflectors, catalyst
Pd	12.0	1555	Pt and Au ores PdSe ore NiS ores	dental and watch alloys, astronomical instruments (mirrors), catalyst
Ag	10.5	961	as free metal or with Cu, Au, and Pb ores sulfide ores	coins, mirrors, jewelry, silverware, electroplating, dental alloys, photography (AgBr)
Cd	8.7	321	impurity in zinc ores CdS $CdCO_3$	low-melting alloys, photoelectric cells, Ni-Cd batteries, dentistry

chemical properties. Scandium has a +3 oxidation state in almost all its compounds. It forms a very stable oxide, Sc_2O_3, and forms halides with the formula ScX_3. The addition of base to $Sc^{3+}(aq)$ produces a white, gelatinous precipitate with the formula $Sc_2O_3 \cdot nH_2O$. Like $Al(OH)_3$, this hydrated Sc_2O_3 is amphoteric. Scandium and its compounds have little technological importance.

Rutile, an ore of titanium.

Titanium

The ground-state electron configuration of titanium is $[Ar]3d^24s^2$. Its most common and stable oxidation state is +4, as in the compounds TiO_2 and $TiCl_4$, which are covalently bonded.

Pure titanium is a lustrous, white metal and is the second most abundant transition metal. It is used to make lightweight alloys that are stable at high temperatures for use in missiles and high-performance aircraft. Titanium is as strong as most steels but 50 percent lighter. It is 60 percent heavier than aluminum but twice as strong. In addition, it has excellent resistance to corrosion.

Table 11-3 Properties of the 5*d* transition series metals

Element	*Density*/g · cm^{-3}	*Melting point*/°C	*Principal sources*	*Uses*
Lu	9.8	1663	monazite	—
Hf	13.3	2227	zirconium minerals	control rods in nuclear reactors, "getter" for oxygen and nitrogen, incandescent lamps
Ta	16.7	2996	niobium ores, especially columbite	pen nibs, balance weights, surgical, dental, and chemical instruments, optical glass
W	19.3	3410 (highest melting point of any metal)	wolframite, (Fe,Mn)WO scheelite, $CaWO_4$	steel toughening, incandescent lamp filaments, electrical contact points, metal cutting tools, catalysis
Re	21.0	3180	rare-earth minerals some sulfide ores	filaments, alloys for electrical contacts, jewelry plating, superconductors
Os	22.61	3045	osmiridium platinum ores	alloy with indium for pen points and machine bearings, catalysis
Ir	22.65 (densest element)	2450	free metal with osmium as alloy platinum ores	pen points, crucibles, platinum hardening
Pt	21.5	1774	natural alloys	jewelry, dentistry, thermometers, electroplating, catalysis
Au	18.9	1065	free metal	jewelry, coins, currency, electrical contacts, dentistry
Hg	13.5	−39 (lowest melting point of any metal)	cinnabar, HgS	barometers, thermometers, Hg arc lamps, silent switches, blasting caps (as fulminate), pharmaceuticals

Pure titanium is difficult to prepare because the metal is very reactive at high temperatures.

The most important ore of titanium is *rutile,* which is primarily TiO_2. Pure titanium metal is produced by first converting TiO_2 to $TiCl_4$, by heating TiO_2 to red heat in the presence of carbon and chlorine. The $TiCl_4$ is reduced to the metal by reacting it with magnesium in an inert atmosphere of argon. Most titanium is used in the production of titanium steels, but TiO_2, which is white when pure, is used as the white pigment in many paints. Titanium tetrachloride is also used to make smoke screens; when it is sprayed into the air it reacts with moisture to produce a dense and persistent white cloud of TiO_2. Titanium dioxide is the only transition-metal compound ranked (46th) in the top 50 industrial chemicals (Appendix A). Some important titanium compounds are given in Table 11-5.

Table 11-4 The common oxidation states of the 3*d* transition metals

Element	*Oxidation state*
Sc	+3
Ti	+4
V	+2, +3, +4, +5
Cr	+2, +3, +6
Mn	+2, +4, +7
Fe	+2, +3
Co	+2, +3
Ni	+2
Cu	+1, +2
Zn	+2

Vanadium

The ground-state electron configuration of vanadium is $[Ar]3d^3 4s^2$. Its maximum oxidation state is +5; the +2, +3, and

Table 11-5 Important titanium compounds

Compound	*Uses*
titanium(IV) carbide, TiC(*s*)	cutting tools
titanium dioxide, $TiO_2(s)$	ceramic colorant, white paints and lacquers, inks and plastics, gemstones
titanium tetrachloride, $TiCl_4(l)$	smoke screens, iridescent glass, artificial pearls

+4 oxidation states are common, with the +2 state being the least common. The oxides VO (black), V_2O_3 (black), VO_2 (dark blue), and V_2O_5 (brick red) are all known. Vanadium(V) oxide is obtained when vanadium is burned in excess oxygen, although some lower oxides are also obtained. Vanadium pentoxide is used as a catalyst in the oxidation of SO_2 to SO_3 in the production of sulfuric acid, as well as several other industrial processes.

Vanadium pentoxide is amphoteric. It dissolves in concentrated bases such as NaOH(*aq*) to produce a colorless solution in which the principal species above pH = 13 is believed to be $VO_4^{3-}(aq)$. As the pH is lowered, the solution turns orange, and at pH = 2, a precipitate occurs, which redissolves at a lower pH to give a pale-yellow solution, in which the principal species is believed to be $VO_2^+(aq)$ (Figure 11-1).

Except for V_2O_5, the compounds of vanadium have limited commercial importance, but vanadium itself is used in alloy steels, particularly *ferrovanadium*. Vanadium also occurs in the +3 and +4 oxidation states in some sea invertebrates. These organisms can accumulate vanadium in their blood at concentrations more than a million times greater than seawater. Some commercially important vanadium compounds are given in Table 11-6.

Figure 11-1 Solutions of V_2O_5. Left, V_2O_5 dissolved in NaOH(*aq*) at pH = 13, where the principal species is $VO_4^{3-}(aq)$. Center left, V_2O_5 dissolved in hydrochloric acid at pH = 0 where the principal species is the vanadyl ion $VO_2^+(aq)$. Center right, V_2O_5 dissolved in hydrochloric acid at pH = 2 with the formation of a precipitate. Right, V_2O_5 dissolved in hydrochloric acid at pH = 4.

Table 11-6 Important vanadium compounds

Compound	*Uses*
vanadium pentoxide, $V_2O_5(s)$	production of sulfuric acid; manufacture of yellow glass; glass ultraviolet-light filters
vanadyl sulfate, $VOSO_4(s)$	blue and green colored glasses and glazes on pottery

11-2 THE +6 OXIDATION STATE OF CHROMIUM AND THE +7 OXIDATION STATE OF MANGANESE ARE STRONGLY OXIDIZING

Chromium

Chromium, with the ground-state electron configuration $[Ar]3d^5 4s^1$, has a maximum oxidation state of +6, although +2 and +3 are common oxidation states. Whereas +4 is the most common oxidation state for titanium and the +5 state of vanadium is only mildly oxidizing, the +6 oxidation state of chromium is strongly oxidizing. The reagents chromium(VI) oxide CrO_3, chromate CrO_4^{2-}, and dichromate $Cr_2O_7^{2-}(aq)$ are strong oxidizing agents. In solutions of CrO_3 with pH greater than 6, the principal species is CrO_4^{2-} (yellow). Between a pH of 2 to 6, the two species $HCrO_4^-$ ($pK_a = 6.51$ at 25°C) and $Cr_2O_7^{2-}$ (orange-red) exist in equilibrium (Figure 11-2):

$$2HCrO_4^-(aq) \rightleftharpoons Cr_2O_7^{2-}(aq) + H_2O(l) \qquad K = 33\ M^{-1} \text{ at } 25°C$$

At pH less than 1, the principal species is H_2CrO_4, chromic acid (deep red).

The dichromate ion in acidic solution is a strong oxidizing agent:

$$14H^+(aq) + Cr_2O_7^{2-}(aq) + 6e^- \rightarrow 2Cr^{3+}(aq) + 7H_2O(l) \qquad E^0 = 1.33\ V$$

whereas the chromate ion in basic solution is a much weaker oxidizing agent:

$$CrO_4^{2-}(aq) + 4H_2O(l) + 3e^- \rightarrow Cr(OH)_3(s) + 5OH^-(aq) \qquad E^0 = -0.13\ V$$

The compound Cr_2O_3 is the most stable oxide of chromium. Solutions of Cr_2O_3 in sulfuric acid are used for electrodepositing chromium in the process of chrome-plating. In contrast to chromium(VI) compounds, the chromium(II) ion is a fairly strong reducing agent:

$$Cr^{3+}(aq) + e^- \rightarrow Cr^{2+}(aq) \qquad E^0 = -0.41\ V$$

Figure 11-2 Left, $Na_2CrO_4(aq)$ in $NaOH(aq)$ at pH = 8, where CrO_4^{2-} is the principal species. Center, a solution from the left with pH adjusted to 4, where the principal species is $Cr_2O_7^{2-}(aq)$. Right, a solution from the center with pH adjusted to 0, where the principal species is $H_2CrO_4(aq)$.

Solutions containing $Cr^{2+}(aq)$ are used often as reducing agents.

The most stable and common oxidation state of chromium is +3, and there are many chromium(III) salts. Some of these are given in Table 11-7.

Manganese

The highest oxidation state of manganese is +7, which is best known in the strongly oxidizing permanganate ion, MnO_4^- (Figure 11-3):

acidic solution

$$MnO_4^-(aq) + 8H^+(aq) + 5e^- \rightarrow Mn^{2+}(aq) + 4H_2O(l) \qquad E^0 = +1.507\ V$$

basic solution

$$MnO_4^-(aq) + 2H_2O(l) + 3e^- \rightarrow MnO_2(s) + 4OH^-(aq) \qquad E^0 = +1.23\ V$$

Figure 11-3 Potassium permanganate, $KMnO_4$, in water is a strong oxidizing agent. Freshly-prepared $KMnO_4(aq)$ is purple; on standing, brown $MnO_2(s)$ precipitates as a result of the decomposition of $KMnO_4(aq)$.

The most important permanganate is potassium permanganate, $KMnO_4$, which is used as an oxidizing agent in industry and medicine, as well as in many general chemistry laboratories. Freshly prepared solutions of potassium permanganate are deep purple, but turn brown on long standing, because permanganate ion oxidizes water to oxygen and is thereby reduced to MnO_2, which is brown. The net reaction is

$$\underset{\text{purple}}{4MnO_4^-(aq)} + 2H_2O(l) \rightarrow \underset{\text{brown}}{4MnO_2(s)} + 3O_2(g) + 4OH^-(aq)$$

Table 11-7 Some important chromium compounds

Compound	*Uses*
chromium(IV) oxide, $CrO_2(s)$	in magnetic recording tapes for better resolution and high-frequency response
chromium(III) oxide, $Cr_2O_3(s)$	in abrasives, refractory materials, and semiconductors; as a green pigment, especially for coloring glass, and when chemical and heat resistance are required
chromium(VI) oxide, $CrO_3(s)$	chromium plating; copper stripping; as a corrosion inhibitor; photography
sodium dichromate, $Na_2Cr_2O_7(s)$	leather tanning; textile manufacture; metal corrosion inhibitor

The reaction is catalyzed by $MnO_2(s)$ and thus is autocatalytic.

Manganese(II) forms soluble salts with most common anions (Table 11-9). For the +3 and +4 oxidation states the most important compounds are the oxides M_2O_3 and MnO_2. Manganese dioxide as the mineral *pyrolusite* is an important ore of manganese and is a source of most manganese compounds.

Some important manganese compounds are given in Table 11-8.

11-3 IRON IS PRODUCED IN A BLAST FURNACE

Iron

With iron we no longer associate the highest oxidation state with the total number of 4*s* and 3*d* electrons. The highest known

Table 11-8 Some important manganese compounds

Compound	*Uses*
manganese(II) chloride, $MnCl_2(s)$	dyeing; disinfectant; purifying natural gas; dry cell manufacture
manganese(IV) oxide, $MnO_2(s)$	manufacture of manganese steel; alkaline batteries and dry cells; amethyst glass; pyrotechnics; painting on porcelain; printing and dyeing textiles; pigment in brick industry
manganese(II) sulfate, $MnSO_4(s)$	dyeing; red glazes on porcelain; boiling oils for varnishes; fertilizers for vines, tobacco; prevention of perosis in poultry
potassium permanganate, $KMnO_4(s)$	oxidizing agent; medical disinfectant (bladder infection); water and air purification

Table 11-9 Some important iron compounds

Compound	*Uses*
iron(III) chloride, $FeCl_3(s)$	treatment of waste water; etching for engraving copper for printed circuitry; feed additive
iron(III) oxide $Fe_2O_3(s)$	metallurgy; paint and rubber pigment; memory cores for computers; magnetic tapes; marine paints; polishing agent for glass, precious metals, and diamonds
iron(0) pentacarbonyl, $Fe(CO)_5(l)$	antiknock agent in motor fuels; manufacture of powdered iron cores in high-frequency coils used in radios and televisions
iron(II) sulfate, $FeSO_4(s)$	flour enrichment; wood preservative; water and sewage treatment; manufacture of ink
iron(III) sulfate, $Fe_2(SO_4)_3(s)$	soil conditioner; disinfectant; etching aluminum; waste water treatment

oxidation state of iron is +6, which is very rare. Only the +2 and +3 oxidation states of iron are common. Iron is the most abundant transition metal, constituting 4.7 percent by mass of the earth's crust. It is the cheapest metal and, in the form of steel, the most useful. Pure iron is a silvery-white, soft metal that rusts rapidly in moist air. It has little use as the pure element but is strengthened greatly by the addition of small amounts of carbon and of various other transition metals. It occurs in nature as hematite, Fe_2O_3, magnetite, Fe_3O_4, siderite, $FeCO_3$, and iron pyrite, FeS_2 (fool's gold) (Figure 11-4).

Figure 11-4 Iron ores. Clockwise from the left: magnetite, Fe_3O_4; siderite, $FeCO_3$; iron pyrite, FeS_2; and hematite, Fe_2O_3.

Millions of tons of iron are produced annually in the United States by the reaction of Fe_2O_3 with coke, which is carried out in a blast furnace. A modern blast furnace is about 100 ft high and 25 ft wide and produces about 5000 tons of iron daily (Figure 11-5). A mixture of iron ore, coke, and limestone ($CaCO_3$) is loaded into the top, and preheated compressed air and oxygen are blown in near the bottom. The reaction of the coke and the oxygen to produce carbon dioxide gives off a great deal of heat, and the temperature in the lower region of a blast furnace is around 1900°C. As the CO_2 rises, it reacts with more coke to produce hot carbon monoxide, which reduces the iron ore to iron. The molten iron metal is denser than the other substances

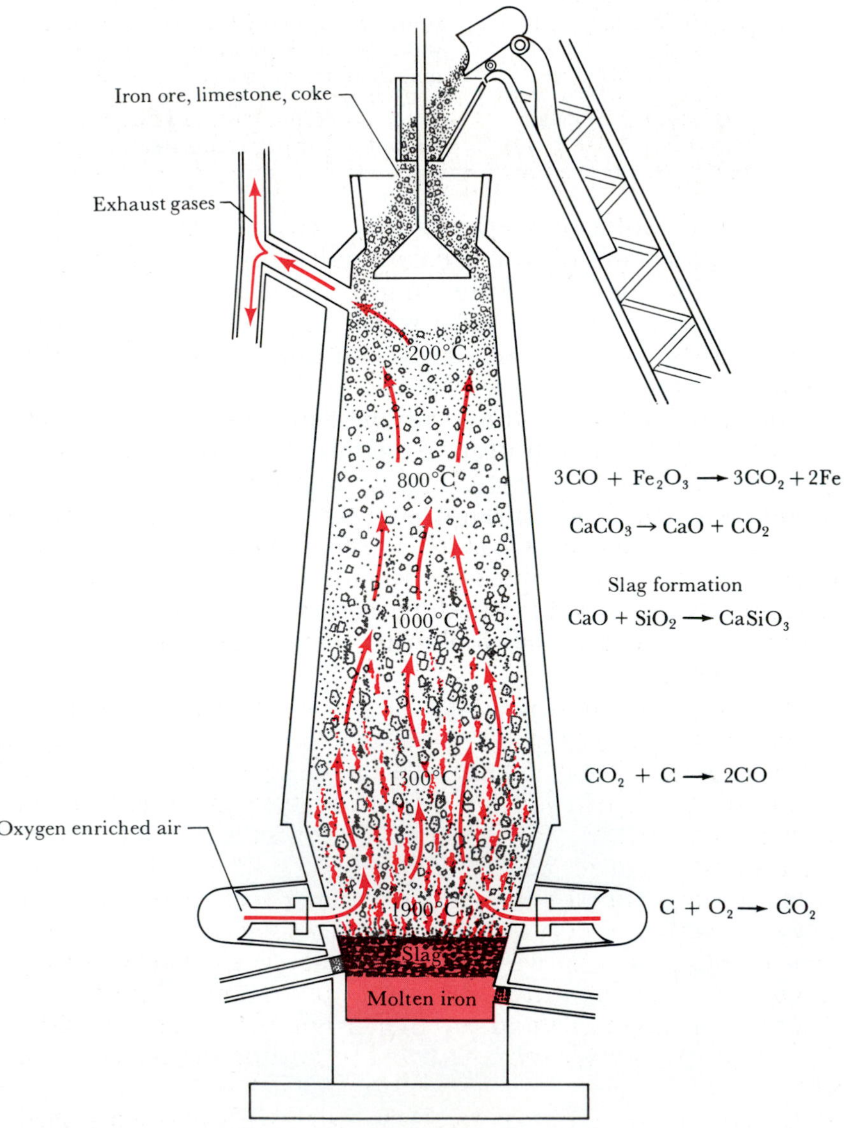

Figure 11-5 A diagram of a blast furnace.

Bethlehem Steel Corp.

These two blast furnaces at Bethlehem Steel's Burns Harbor plant have a combined total production of over 10,000 tons of iron daily. Ore, limestone, and coke are fed directly into the top of the furnaces by an automatic conveyor system.

Bethlehem Steel Corp.

Figure 11-6 Molten iron being poured into a basic oxygen furnace. Most steel is produced by a process called the basic oxygen process. A typical basic oxygen furnace is charged with about 200 tons of molten pig iron, 100 tons of scrap iron, and 20 tons of limestone (to form a slag). A stream of hot oxygen is blown through the molten mixture, where the impurities are oxidized and blown out of the iron. High-quality steel is produced in an hour or less.

and drops to the bottom, where it can be drained off to form ingots of what is called *pig iron.*

The function of the limestone is to remove the sand and gravel that normally occur with iron ore. The intense heat decomposes the limestone to CaO and CO_2. The CaO(*s*) combines with the sand and gravel (both of which are primarily silicon dioxide) to form calcium silicate, $CaSiO_3(s)$. The molten calcium silicate, called *slag*, floats on top of the molten iron and is drained off periodically. It is used in building materials, such as cement and concrete aggregate, rock-wool insulation, and cinder block, and as railroad ballast.

Pig iron contains about 4 or 5 percent carbon together with lesser amounts of silicon, manganese, phosphorus, and sulfur. It is brittle, difficult to weld, and not strong enough for structural applications. To be useful, pig iron must be converted to steel, which is an alloy of iron with small but definite amounts of other metals and between 0.1 and 1.5 percent carbon. Steel is made from pig iron in several different processes, all of which use oxygen to oxidize most of the impurities. One such process is the *basic oxygen process,* in which hot, pure O_2 gas is blown through molten pig iron (Figure 11-6). The oxidation of carbon and phosphorus is complete in less than 1 hour. The desired carbon content of the steel is then achieved by adding high-carbon steel alloy.

There are two types of steels, carbon steels and alloy steels. Both types contain carbon, but carbon steels contain essentially no other metals besides iron. About 90 percent of all steel produced is carbon steel. Carbon steel that contains less than 0.2 percent carbon is called *mild steel.* Mild steels are malleable and ductile and are used where load-bearing ability is not a consideration. *Medium steels,* which contain 0.2 to 0.6 percent carbon, are used for such structural materials as beams and girders and for railroad equipment. *High-carbon steels* contain 0.8 to 1.5 percent carbon and are used to make drill bits, knives, and other tools in which hardness is important.

▪ Chromium improves hardness and resistance to corrosion.
Tungsten and molybdenum increase heat resistance.
Nickel adds toughness, as in armor plating.
Vanadium adds springiness.
Manganese improves resistance to wear.

Alloy steels contain other metals in small amounts. Different metals give different properties to steels. The alloy steels called *stainless steels* contain high percentages of chromium and nickel. Stainless steels resist corrosion and are used for cutlery and hospital equipment. The most common stainless steel contains 18 percent chromium and 8 percent nickel.

11-4 BILLIONS OF DOLLARS ARE SPENT EACH YEAR TO PROTECT METALS FROM CORROSION

We are all familiar with corrosion, the best-known example of which is the rusting of iron and steel. Rust is iron(III) oxide, and the corrosion of iron proceeds by air oxidation of the iron:

$$4Fe(s) + 3O_2(g) \xrightarrow{H_2O(l)} \underset{\text{rust}}{2Fe_2O_3(s)}$$

Most metals, when exposed to air, develop an oxide film. In some cases this film is very thin and protects the metal, and the metal maintains its luster. However, in other cases (such as iron) corrosion can completely destroy the metal. Corrosion of metals is promoted by moisture, acidic oxides, and anions, such as Cl^-, which form chloro complexes.

Corrosion involves electron-transfer reactions between different sections of the same piece of metal or between two dissimilar metals in electrical contact with each other. One metal piece acts as the anode, and the other acts as the cathode. For example, iron in contact with air and moisture corrodes according to the mechanism sketched in Figure 11-7. The anodic process is

$$2Fe(s) + 4OH^-(aq) \rightarrow 2Fe(OH)_2(s) + 4e^-$$

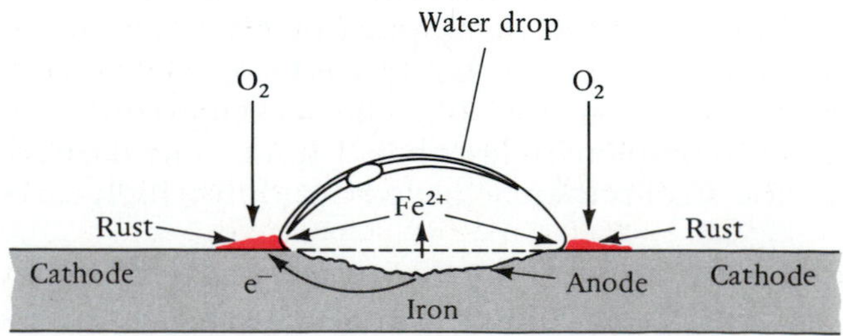

Figure 11-7 A drop of water on an iron surface can act as a corrosion center. The iron is oxidized by oxygen from the air. Moisture is necessary for corrosion because the mechanism involves the formation of dissolved $Fe^{2+}(aq)$ ions.

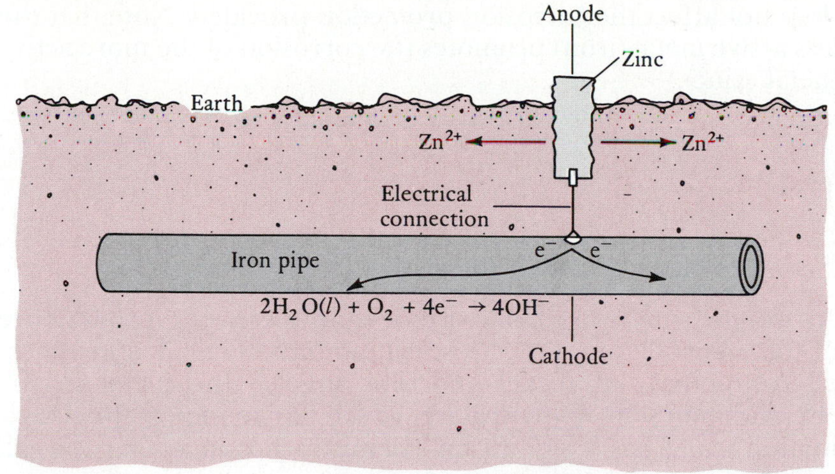

Figure 11-8 Protection of an iron pipe from corrosion with a sacrificial zinc anode. Zinc is a stronger reducing agent than iron and thus is preferentially oxidized.

and the cathodic process is

$$2H_2O(l) + O_2(g) + 4e^- \rightarrow 4OH^-(aq)$$

where the $O_2(g)$ comes from the air. The iron (II) hydroxide formed is rapidly air-oxidized in the presence of water to iron(III) hydroxide:

$$4Fe(OH)_2(s) + O_2(g) + 2H_2O(l) \rightarrow 4Fe(OH)_3(s)$$

which in turn converts spontaneously to iron(III) oxide:

$$2Fe(OH)_3(s) \rightarrow Fe_2O_3 \cdot 3H_2O(s)$$

The corrosion of aluminum in air is not so pronounced as that of iron because the Al_2O_3 film that forms is tough and adherent and impervious to oxygen. The same is true for chromium and nickel.

The simplest method of corrosion prevention is to provide a protective layer of paint or of a corrosion-resistant metal, such as chromium or nickel. The weakness of such methods is that any scratch or crack in the protective layer exposes the metal surface. The exposed surface, even though small in area, can act as an anode in conjunction with other exposed metal parts, which act as cathodes. This combination then leads to corrosion of the metal under the no-longer-protective layer.

Another anticorrosion technique uses a replaceable *sacrificial anode,* which is a piece of metal electrically connected to a less active metal (Figure 11-8). The more active metal is the stronger reducing agent and is thus preferentially oxidized; oxygen is reduced on the surface of the less active metal. Sacrificial anodes are used to protect water pipes and ship propellers. This method is also the electrochemical basis of galvanization, in which iron is protected from corrosion by a zinc coating, a process used in the manufacture of automobile bodies. A crack in the zinc coating does not affect the corrosion protection provided. Note that the less active metal (iron) promotes the corrosion of the more active metal (zinc).

11-5 THE +2 OXIDATION STATE IS THE MOST IMPORTANT OXIDATION STATE FOR COBALT, NICKEL, COPPER, AND ZINC

As we go from iron to zinc, there is an increasing prominence of the +2 oxidation state. Most compounds of cobalt and almost all compounds of nickel and zinc involve the metal in the +2 oxidation state. Only copper, which has an important +1 oxidation state, has a significant chemistry not involving the +2 state.

Cobalt

Cobalt is a fairly rare element and is usually found associated with nickel in nature. It is a hard, bluish-white metal that is used in the manufacture of high-temperature alloys and permanent magnets (Alnico). The pure metal is relatively unreactive, and dissolves only slowly in dilute mineral acids. When cobalt is burned in oxygen, a mixture of CoO and Co_3O_4 is obtained. Most simple cobalt salts involve Co(II), and many of these are pink or red. The species $Co^{3+}(aq)$ is a strong oxidizing agent:

$$Co^{3+}(aq) + e^- \rightarrow Co^{2+}(aq) \qquad E^0 = +1.84 \text{ V}$$

and oxidizes $H_2O(l)$.

Some important cobalt compounds are given in Table 11-10.

Nickel

Nickel is the twenty-second most abundant element in the earth's crust and occurs in a variety of sulfide ores, the most important deposit being found in the Sudbury basin of Ontario, Canada. The metal is obtained by roasting the ore to obtain NiO, and then reducing with hydrogen or carbon. Nickel is a silvery metal that takes a beautiful high polish. The bulk, pure metal is highly corrosion-resistant and is often used as a protective coating. It is used in a number of magnetic alloys, and in the alloy Monel, which is used to handle fluorine and other reactive fluorine compounds. As a powder, nickel is used as a catalyst for the hydrogenation of vegetable oils. Nickel is more reactive than cobalt, and dissolves readily in dilute acids. The aqueous solution chemistry of nickel involves primarily the species $Ni^{2+}(aq)$. Some commercially important nickel compounds are given in Table 11-11.

Table 11-10 Some important cobalt compounds

Compound	*Uses*
cobalt(II) chloride, $CoCl_2(s)$	absorbent for ammonia; gas masks; hygrometers; solid lubricant; electroplating; invisible inks
cobalt(II) oxide, $CoO(s)$	in glass and ceramic coloring and decolorization
cobalt(II) phosphate, $Co_3(PO_4)_2(s)$	lavender pigment in paints and ceramics
cobalt(II) sulfate, $CoSO_4(s)$	storage batteries; electrode manufacture; cobalt electroplating baths; in ceramics, enamels, glazes to prevent discoloring
cobalt(III) fluoride, $CoF_3(s)$	fluorinating agent

Copper

Copper is slightly less abundant than nickel. Copper generally occurs as various sulfides, although in some ores copper is present in the form of sulfates, carbonates, and other oxygen containing compounds. Deposits of the free metal are very rare, being found only in Michigan. Most copper-containing deposits have a copper content of less than 1 percent, but some richer deposits have up to 4 percent copper. Copper ores contain other metals and semimetals such as selenium and tellurium, which are important by-products of copper production. Some important copper minerals are *chalcocite* (Cu_2S), *chalcopyrite* ($CuFeS_2$), and *malachite* ($CuCO_3 \cdot Cu(OH)_2$) (Figure 11-9).

▪ The total world output of copper metal in 1982 was about 8 million metric tons, with U.S. production accounting for 14% of the world total.

Most of the world copper production is *pyrometallurgical,* that is, based on the smelting of sulfide ores. Before smelting, the copper minerals in the ore are crushed to fine particles and concentrated. The ore concentrate and a mixture of silica, SiO_2, iron oxide, FeO, and lime, CaO, are added to a reverberatory furnace from above. Two molten products are formed, which separate into two layers and are drained separately. One layer, the slag,

Table 11-11 Some important nickel compounds

Compound	*Uses*
nickel(0) carbonyl, $Ni(CO)_4(l)$	manufacture of high-purity nickel powder and pellets; nickel coatings on steel
nickel(II) cyanide, $Ni(CN)_2(s)$	metallurgy; electroplating
nickel(II) chloride, $NiCl_2(s)$	nickel-plating; absorbent for NH_3 in gas masks
nickel(II) sulfate, $NiSO_4(s)$	nickel-plating; mordant in dyeing and printing fabrics; preparation of catalysts

Figure 11-9 The major copper ores. From left to right: chalcopyrite, $CuFeS_2$; malachite, $CuCO_3 \cdot Cu(OH)_2$; and chalcocite, Cu_2S.

consists mainly of iron silicates and may be reused in the furnace. The other layer, the *matte,* consists primarily of copper and iron sulfides.

The matte is converted to a form of copper metal called *blister copper* in a converter similar to the basic oxygen process used in making steel. The molten matte is poured into a furnace and air is injected along the sides. In the first stage iron is removed by adding silica and injecting air for several hours. The iron compounds formed (slag) are insoluble in the molten matte. The slag is lighter than the matte and is skimmed off.

After the removal of the slag, air is again injected to produce metallic copper. The reactions that occur at this stage are

$$2Cu_2S(l) + 3O_2(g) \rightarrow 2Cu_2O(l) + 2SO_2(g)$$
$$2Cu_2O(l) + Cu_2S(s) \rightarrow 6Cu(l) + SO_2(g)$$

The copper metal is insoluble in the matte and sinks to the bottom of the furnace. The blister copper produced is 96 to 99.5 percent copper, with oxygen being the main impurity. Blister copper cannot be cast and is refined before final purification by electrolysis. Oxygen, which is present in Cu_2O, is removed by reaction with hydrogen and carbon monoxide in a refining furnace. The copper metal is purified by electrolysis (Figure 11-10).

Copper is a soft, ductile metal with a distinctive reddish color. Copper is important because of its use as an electrical conductor. Silver is the only metal that is a better conductor than copper, but the price of silver precludes its widespread use. Copper is reddish and takes on a bright metallic luster.

Ray Manley Photography

Figure 11-10 Copper metal is purified by electrorefining. A current is passed between a pure copper metal cathode and an impure copper metal anode that are immersed in a solution containing Cu(II). Copper metal is oxidized to Cu(II) at the anode and Cu(II) is reduced to pure copper metal at the cathode. Note the refined copper electrodes that have been removed from a vat in the center of the photo.

Brass is an alloy of copper with zinc, and bronze is an alloy of copper with tin. Brass and bronze are among the earliest known alloys. Bronze usually contains from 5 to 10 percent tin and is very resistant to corrosion. It is used for casting, marine equipment, fine arts work, and spark-resistant tools. Yellow brasses contain about 35 percent zinc and have good ductility and high strength. Brass is used for piping, hose nozzles, marine equipment, and jewelry and in the fine arts.

Although copper is fairly unreactive, its surface turns green after long exposure to the atmosphere. The green patina (Figure 11-11) is due to the surface formation of copper hydroxo carbonate and hydroxo sulfate. Copper does not replace hydrogen from dilute acids, because

$$Cu^{2+}(aq) + 2e^- \rightarrow Cu(s) \qquad E^0 = +0.34 \text{ V}$$

but it reacts with oxidizing acids such as dilute nitric acid or hot concentrated sulfuric acid:

$$3Cu(s) + 8HNO_3(aq) \rightarrow 3Cu(NO_3)_2(aq) + 2NO(g) + 4H_2O(l)$$
$$Cu(s) + 2H_2SO_4(\text{conc}) \rightarrow CuSO_4(aq) + 2H_2O(l) + SO_2(g)$$

Most compounds of copper involve Cu(II). Copper(I) salts are

Peter Arnold

Figure 11-11 The statue of Liberty has turned green as a result of the corrosion of copper metal. A major restoration project of the statue's surface began in 1984.

often colorless and only slightly soluble in water. The $Cu^{+}(aq)$ ion is unstable and disproportionates according to

$$2Cu^{+}(aq) \rightarrow Cu(s) + Cu^{2+}(aq) \qquad K = 1.8 \times 10^{6}\ M^{-1}$$

Many copper(II) salts are blue or bluish-green. The most common copper(II) salt is copper(II) sulfate pentahydrate, $CuSO_4 \cdot 5H_2O$, which occurs as beautiful blue crystals. When the crystals are heated gently, the waters of hydration are driven off to produce anhydrous copper(II) sulfate, $CuSO_4$, which is a white powder (Figure 11-12). Some important copper compounds are given in Table 11-12.

Zinc

▪ Smithsonite is named after James Smithson, the founder of the Smithsonian Institution.

Zinc is widely distributed in nature and is about as abundant as copper. Its principal ores are *sphalerite* (ZnS), or zinc blende, and *smithsonite* ($ZnCO_3$), from which zinc is obtained by roasting and reduction of the resultant ZnO with carbon. Zinc is a shiny white metal with a bluish-gray luster.

The 3*d* subshell of zinc is completely filled, and zinc behaves more like a Group 2 metal than like a transition metal. Metallic zinc is a strong reducing agent. It dissolves readily in dilute acids

Table 11-12 Some important copper compounds

Compound	*Uses*
copper(II) arsenite, $Cu(AsO_2)_2(s)$ (Scheele's green)	pigment; wood preservative; insecticide, fungicide, rodenticide, mosquito control
copper(II) chloride, $CuCl_2(s)$	pyrotechnics; wood preservative; photography; electroplating; in the petroleum industry as a deodorizing, desulfurizing, and purifying agent, catalyst
copper(II) hydroxide, $Cu(OH)_2(s)$	fungicides; insecticides; feed additive; staining paper; Bordeaux mixture
copper(II) nitrate, $Cu(NO_3)_2(s)$	light-sensitive paper; insecticide for vines; burnishing iron and copper black; aluminum brighteners; wood preservative; catalyst component in rocket solid fuels
copper(I) oxide, $Cu_2O(s)$	antifouling points for wood and steel exposed to seawater; fungicide; porcelain red glaze; red glass
copper(II) sulfate, $CuSO_4 \cdot 5H_2O(s)$	soil and feed additive; germicide; leather industry; medicine; petroleum and rubber industry; detection and removal of trace amounts of water from alcohols; laundry and metal-marking inks

Figure 11-12 Copper(II) sulfate pentahydrate, $CuSO_4 \cdot 5H_2O(s)$, is blue, and copper(II) sulfate, $CuSO_4$, is white.

Table 11-13 Some important zinc compounds

Compound	*Uses*
zinc(II) carbonate, $ZnCO_3(s)$	white pigment; porcelain, pottery and rubber manufacture; astringent and antiseptic
zinc(II) chloride, $ZnCl_2(s)$	deodorants; disinfectants; fire-proofing and preserving wood; adhesives; dental cements; taxidermist fluid; artificial silk; parchment paper
zinc(II) oxide, $ZnO(s)$	ointment; pigment and mold inhibitor in paints; floor tile; cosmetics; color photography; dental cements; automobile tires
zinc(II) sulfate, $ZnSO_4(s)$	soil additive; wood and skin preservative; clarifying glue
zinc(II) sulfide, $ZnS(s)$	white pigment for low-gloss paints; oilcloths, linoleum, leather, dental rubber; in X-ray screens and luminous dials of watches

and combines with oxygen, sulfur, phosphorus, and the halogens upon being heated. The only important oxidation state of zinc is +2, and zinc(II) salts are colorless, unless color is imparted by the anion. Some important zinc(II) salts are given in Table 11-13.

11-6 GOLD, SILVER, AND MERCURY WERE KNOWN IN THE ANCIENT WORLD

Gold is a very dense, soft, yellow metal with a high luster (Figure 11-13). It is found in nature as the free element and in tellurides. It occurs in veins and alluvial deposits and is often separated from rocks and other minerals by sluicing or panning. Over two thirds of the gold produced by the noncommunist world comes from South Africa. In many mining operations, about 5 g of gold is recovered from 1 ton of rock. Gold is very unreactive and has a remarkable resistance to corrosion. Pure gold is soft and often alloyed to make it harder. The amount of gold in an alloy is expressed in *karats*. Pure gold is 24 karat. Coinage gold is 22 karat, or $(22/24) \times 100 = 92$ percent. White gold, which is used in jewelry, is usually an alloy of gold and nickel. In addition to its use in jewlery and as a world monetary standard, gold is an excellent conductor of electricity and is used in microelectronic devices. It is also used extensively in dentistry and medicine.

Gold is extracted from ores by reaction with sodium cyanide, NaCN and oxygen:

$$4Au(s) + 8CN^-(aq) + O_2(g) + 2H_2O(l) \rightarrow 4Au(CN)_2^-(aq) + 4OH^-(aq)$$

Echo Bay Mines

Figure 11-13 Gold bars from Echo Bay Mines in Edmonton, Alberta.

The gold is recovered from the $Au(CN)_2^-$ either by the replacement reaction:

$$2Au(CN)_2^-(aq) + Zn(dust) \rightarrow Zn(CN)_4^{2-}(aq) + 2Au(s)$$

or by electrolysis. Some important gold compounds are listed in Table 11-14.

Silver is a lustrous, white metal whose ductility and malleability are exceeded only by gold and palladium. Pure silver has the highest electrical conductivity of all metals. Most of the silver produced nowadays is a by-product of the production of other metals such as copper, lead, and zinc. Silver is used in jewelry, silverware, high-capacity batteries, coinage, and photography. About a third of the silver produced is used in photography.

The sensitivity of silver halide crystals to light is the basis of the photographic process. Photographic film consists of very small silver halide crystals dispersed uniformly throughout some gelatinous substance, which itself spread over a transparent film. Such a dispersion is known as an *emulsion*. Silver bromide is commonly used, but silver iodide is used for fast film. When a photon of light falls upon a silver halide crystal, a halide ion loses an electron, which eventually migrates to the surface of the crystal and reduces a silver ion:

$$X^-(s) + h\nu \rightarrow X(s) + e^-$$

$$Ag^+(s) + e^- \rightarrow Ag(s)$$

The silver, being very finely divided, appears black, and the pattern of silver atoms throughout the emulsion at this stage is called a *latent image*. The latent image is now intensified by a *developer*, which is a mild reducing agent that reduces more silver ions. A common developer is hydroquinone, HO–C_6H_4–OH, and the half-reactions are

$$AgX(s) + e^- \rightarrow Ag(s) + X^-$$

$$HO\text{–}C_6H_4\text{–}OH \longrightarrow O\text{=}C_6H_4\text{=}O + 2H^+ + 2e^-$$

Table 11-14 Some important gold compounds

Compound	*Uses*
aurothioglycanide, $C_8H_8AuNOS(s)$	antiarthritic agent
gold(I) stannate, $Au_2SnO_2(s)$	manufacture of ruby glass, colored enamels, and porcelain
tetrachloroauric(III) acid, $HAuCl_4(s)$	photography; gold plating; gilding glass and porcelain

Table 11-15 Some important silver compounds

Compound	*Uses*
silver iodide, $AgI(s)$	dispersed in clouds to induce rain; photography (fast film)
silver nitrate, $AgNO_3(s)$	photography; manufacture of mirrors; silver-plating; hair-darkening agent; eye drops for newborn infants
silver oxide, $Ag_2O(s)$	purification of drinking water; polishing and coloring glass (yellow)
silver bromide, $AgBr(s)$	photography; photosensitive lenses

Cinnabar, $HgS(s)$, and mercury, $Hg(l)$.

The overall reaction is catalyzed by the silver atoms of the latent image, and so the reduction occurs preferentially in those areas of the emulsion that received the most light.

The film is *fixed* by removing all the remaining silver halide to prevent further reduction. This is commonly done with a solution of sodium thiosulfate (*hypo*), which reacts with the silver halide according to

$$AgX(s) + 2Na_2S_2O_3(aq) \rightarrow Na_3[Ag(S_2O_3)_2](aq) + NaX(aq)$$

The resulting image is called a *negative* because the area that corresponds to the most light appears the darkest. A positive print is obtained by allowing light to pass through the negative onto printing paper, which is developed and fixed by the above process. Some important silver compounds are given in Table 11-15.

Like gold and silver, mercury has been known for thousands of years. Mercury is the only metal that is a liquid at 0°C and used to be called quicksilver. The principal ore of mercury is *cinnabar,* HgS, which was widely used in the ancient world as a vermilion pigment. The most extensive and richest deposits of cinnabar occur in the Almadén region of Spain, the world's largest producer of mercury. The metal is easily recovered from its ore by roasting:

▪ The symbol Hg is derived from the Latin *hydragyrum,* meaning liquid silver.

$$HgS(s) + O_2(g) \rightarrow Hg(l) + SO_2(g)$$

and the mercury is purified by distillation.

Mercury is not very reactive. On being heated, it reacts with oxygen, sulfur, and the halogens, but not with nitrogen, phosphorus, hydrogen, or carbon. When mercury is heated in air to around 300°C, it reacts with oxygen to produce the bright orange-red mercury(II) oxide. When HgO is heated to about 400°C, it decomposes according to

$$2HgO(s) \xrightarrow{400^\circ C} 2Hg(l) + O_2(g)$$

The property of absorbing oxygen from air and regenerating it as pure oxygen played a key role in the pioneering work of

Lavoisier and Priestley on oxygen. Like copper, mercury does not replace hydrogen from acids, but it does react with oxidizing acids such as dilute nitric acid or hot concentrated sulfuric acid:

$$3Hg(l) + 8HNO_3(aq) \rightarrow 3Hg(NO_3)_2(aq) + 2NO(g) + 4H_2O(l)$$

$$Hg(l) + 2H_2SO_4(aq) \rightarrow HgSO_4(aq) + SO_2(aq) + 2H_2O(l)$$

Mercury compounds occur as Hg(I) or Hg(II), with Hg(II) being the more common. Except for the nitrate, acetate, and perchlorate, Hg(I) salts are insoluble. A notable feature of Hg(I) salts is that they consist of the diatomic Hg_2^{2+} ion. Many of the salts of Hg(I) and Hg(II) are covalently bonded.

▪ Metallic mercury is also very poisonous, and its appreciable vapor pressure poses a serious danger of inhalation.

Mercury compounds are very poisonous and have been used in insecticides, fungicides, rodenticides, antifouling paints, and disinfectants. In the last century, mercury compounds were used in the production of felt for hats, and the workers involved suffered from a severe nervous disorder called "hatter's shakes," which led to the expression "mad as a hatter." The discharge of mercury-containing industrial wastes into rivers, lakes, and the oceans caused serious environmental problems, culminating in the Minamata disaster in Japan in 1952, when over 50 people died of mercury poisoning. The mercury effluent was converted to the organomercury compounds by certain sedimentary bacteria and entered the marine food chain, and became concentrated in fish, which was the main diet of the fishing village of Minamata. Since then, the disposal of mercury wastes has been regulated and mercury levels in food are constantly monitored. In 1972 more than 90 nations agreed on an international ban on the dumping of mercury wastes. Some important mercury compounds are listed in Table 11-16.

The transition metals have a rich and fascinating chemistry involving numerous ligands and different oxidation states. The coordination chemistry of the transition metals is described in Chapter 23 of the text.

When mercury(II) oxide is heated, it decomposes into elemental mercury and oxygen gas. The red compound shown here is mercury(II) oxide. The elemental liquid mercury has condensed on the walls of the test tube. This reaction was one of the earliest methods used to produce oxygen in the laboratory. It is no longer used, however, because of the toxicity of the mercury vapor produced.

Table 11-16 Some important mercury compounds

Compound	*Uses*
mercury(I) chloride, $Hg_2Cl_2(s)$ (calomel)	fungicide; control of root maggots on cabbage and onions; calomel electrodes
mercury(II) chloride, $HgCl_2(s)$	preservative for wood and anatomical specimens; embalming agent; photographic intensifier
mercury(II) oxide, red, $HgO(s)$	marine paints; porcelain pigments; anode material in mercury batteries
mercury(II) sulfide, red, $HgS(s)$	coloring plastics and papers; sealing wax

TERMS YOU SHOULD KNOW

fool's gold
blast furnace
pig iron
slag
basic oxygen process
sacrificial anode
pyrometallurgical
matte
blister copper
karat
latent image
developer
fixer
hypo

QUESTIONS

11-1. Which metal has the highest melting point?

11-2. Which are the two densest metals?

11-3. Which is the most abundant transition metal?

11-4. Which are the only two 3*d* transition metals that do not readily form divalent ions in aqueous solution?

11-5. Describe how titanium metal is produced from its ore.

11-6. Give the highest oxidation states for Sc, Ti, V, Cr, and Mn.

11-7. Name a catalyst in the production of sulfuric acid by the contact process.

11-8. The chief ore of chromium is *chromite* ($FeCr_2O_4$). Chromium of high purity can be obtained from chromite by oxidizing chromium(III) to chromium(VI) in the form of sodium dichromate, $Na_2Cr_2O_7$, and then reducing it with carbon according to

$$Na_2Cr_2O_7(s) + 2C(s) \rightarrow Cr_2O_3(s) + Na_2CO_3(s) + CO(g)$$

The oxide is then reduced with aluminum by the thermite reaction:

$$Cr_2O_3(s) + 2Al(s) \rightarrow Al_2O_3(s) + 2Cr(s)$$

If an ore is 65.0 percent chromite, then how many grams of pure chromium can be obtained from 100 grams of ore?

11-9. A small quantity of $Na_2Cr_2O_7$ is often added to water stored in steel drums used in some passive home solar heating system. The dichromate acts as a corrosion inhibitor by forming an impervious layer of $Cr_2O_3(s)$ on the iron surface. Write a balanced chemical equation for the process in which Cr_2O_3 is formed.

11-10. A chromous bubbler is used to remove traces of oxygen from various gases, for example, from tank nitrogen. The bubbler solution is prepared by reducing $Cr^{3+}(aq)$ to $Cr^{2+}(aq)$ with excess zinc metal. Chromium(II) rapidly reduces O_2 to water and forms chromium(III), which is then reduced back to Cr(II) by the zinc. Write balanced chemical equations for the various chemical reactions involved in the operation of the bubbler.

11-11. Why are solutions of potassium permanganate stored in dark bottles? What is the reaction?

11-12. The reaction

$$MnO_2(s) + 4HCl(g) \rightarrow MnCl_2(aq) + Cl_2(g) + 2H_2O(l)$$

is often used to generate small quantities of $Cl_2(g)$ in the laboratory. How many grams of $Cl_2(g)$ can be generated by reacting 4.25 g MnO_2 with an excess of $HCl(aq)$?

11-13. Use VSEPR theory to predict the shapes of (a) $TiCl_4$, (b) VF_5, (c) CrO_4^{2-} and (d) MnO_4^-.

11-14. Describe the principal reactions that take place in a blast furnace.

11-15. Describe the basic oxygen process.

11-16. Suppose that an iron ore consists of 50 percent Fe_2O_3 and 50 percent SiO_2. How many metric tons of iron and slag will be produced from 10,000 metric tons of ore?

11-17. Why isn't the corrosion of aluminum as serious a problem as the corrosion of iron?

11-18. Describe how a sacrificial anode works.

11-19. Why must solutions of $Co^{3+}(aq)$ be prepared freshly?

11-20. Which two elements are important byproducts of copper production?

11-21. A copper ore consists of 2.65 percent chalcopyrite. How many tons of ore must be processed to obtain one metric ton of copper?

11-22. When copper reacts with dilute nitric acid, NO(*g*) is evolved. Although NO(*g*) is colorless, it appears as though a brown-red gas is evolved. Explain.

11-23. What is the percentage of gold in 14-karat gold?

11-24. Describe the photographic process, including the roles of the latent image, the developer, and the fixer.

TOP 50 CHEMICALS (1983)

Rank		*Billions of lb*	*Rank*		*Billions of lb*
1	Sulfuric acid	69.45	26	Formaldehyde	5.40
2	Nitrogen	42.03	27	Hydrochloric acid	5.22
3	Lime	28.80	28	Ethylene glycol	4.46
4	Oxygen	28.73	29	*p*-Xylene	4.11
5	Ethylene	28.59	30	Ammonium sulfate	3.94
6	Ammonia	27.37	31	Cumene	3.30
7	Sodium hydroxide	20.46	32	Potash	2.87
8	Chlorine	19.92	33	Acetic acid	2.79
9	Phosphoric acid	19.90	34	Phenol	2.61
10	Sodium carbonate	16.93	35	Carbon black	2.50
11	Nitric acid	14.75	36	Butadiene	2.31
12	Propylene	13.98	37	Aluminum sulfate	2.29
13	Ammonium nitrate	13.24	38	Acrylonitrile	2.15
14	Urea	11.54	39	Vinyl acetate	1.96
15	Ethylene dichloride	11.25	40	Calcium chloride	1.88
16	Benzene	9.48	41	Acetone	1.87
17	Ethylbenzene	7.86	42	Sodium sulfate	1.71
18	Carbon dioxide	7.15	43	Cyclohexane	1.69
19	Toluene	7.12	44	Propylene oxide	1.58
20	Styrene	6.99	45	Titanium dioxide	1.51
21	Vinyl chloride	6.95	46	Sodium silicate	1.45
22	Methanol	6.62	47	Adipic acid	1.42
23	Terephthalic acid	5.69	48	Sodium tripolyphosphate	1.34
24	Ethylene oxide	5.58	49	Isopropyl alcohol	1.21
25	Xylene	5.57	50	Ethanol	1.10

Bureau of the Census, Bureau of Mines, International Trade Commission, C&EN estimates.

ANSWERS

Chapter 1

1-1. See frontispiece.
1-2. 10; 14
1-3. The principal quantum number of the valence *f* subshell is two less than the principal quantum number of the outermost ground-state electrons in the neutral atom.
1-4. See Table 1-2.
1-5. a, d, e, and f are metals; c, g, and h are semimetals; b is a nonmetal.
1-6. Se: $1s^22s^22p^63s^23p^64s^23d^{10}4p^4$;
Cs: $1s^22s^22p^63s^23p^64s^23d^{10}4p^65s^24d^{10}5p^66s^1$;
Sb: $1s^22s^22p^63s^23p^64s^23d^{10}4p^65s^24d^{10}5p^3$;
In: $1s^22s^22p^63s^23p^64s^23d^{10}4p^65s^24d^{10}5p^1$
1-7. a is an *s*-block element; c, d, e, and f are *p*-block elements; b is a *d*-block element; g and h are *f*-block elements.
1-8. a, b, and d are not expected to exist.
1-9. Group 3 or 5
1-10. See Figures 1-1, 1-4, and 1-6.
1-11. a and c decrease; b, d, and e increase.
1-12. The van der Waals attraction increases with increasing atomic size.
1-13. Third-row elements can accommodate more than eight electrons in their valence shells.
1-14. The pairs of elements Li–Mg, Be–Al, and B–Si have many similar chemical properties.
1-15. The oxidation state that is two less than the maximum possible becomes increasingly stable on going down a group.

Chapter 2

2-1. 1.0079
2-2. Add trace amounts of T_2O or THO and follow its movement by measuring the radioactivity at various locations.
2-3. 1.07×10^9 disintegrations per second
2-4. 686 disintegrations per second
2-5. (a) $Fe_2O_3(s) + 3H_2(g) \rightarrow 2Fe(s) + 3H_2O(g)$
(b) $LiH(s) + H_2O(l) \rightarrow LiOH(aq) + H_2(g)$
(c) $Mg(s) + H_2(g) \rightarrow MgH_2(s)$
(d) $2K(s) + H_2(g) \rightarrow 2KH(s)$
(e) $H_2(g) + H_2C{=}CH_2(g) \xrightarrow{Pt} CH_3CH_3(g)$

2-6. (a) $Zn(s) + 2HBr(aq) \rightarrow ZnBr_2(aq) + H_2(g)$
(b) $C(s) + H_2O(g) \xrightarrow[1000°C]{Fe} CO(g) + H_2(g)$
(c) $3D_2(g) + N_2(g) \xrightarrow{Fe/Mo} 2ND_3(g)$
(d) $2Li(s) + 2D_2O(l) \rightarrow 2LiOD(aq) + D_2(g)$
(e) $W(s) + 3H_2O(g) \rightarrow WO_3(s) + 3H_2(g)$
2-7. $Zr(s) + 2H_2O(g) \rightarrow ZrO_2(s) + 2H_2(g)$
2-8. $C_3H_8(g) + 3H_2O(g) \xrightarrow[1000°C]{Ni} 3CO(g) + 7H_2(g)$
2-9. $2Li(s) + 2CH_3CH_2OH(l) \rightarrow 2LiOCH_2CH_3(aq) + H_2(g)$
2-10. Hydrogen forms explosive mixtures with air over a wide composition range (2 to 98% H_2). Due to its small mass, it also effuses and diffuses readily.
2-11. 1.32 g
2-12. 1.17 g H_2 from Fe versus 1.00 g H_2 from Zn
2-13. 1.39×10^5 L
2-14. 120 $kJ \cdot g^{-1}$
2-15. 0.990 V
2-16. 1.23 V

Chapter 3

3-1. Because they react with air.
3-2. The ionization energy decreases with increasing atomic number.
3-3. By the electrolysis of NaCl(l)
3-4. NH_3 (recovered), NaCl, $CaCO_3$, and H_2O (see Section 3-5)
3-5. (a) $2Na(s) + 2H_2O(l) \rightarrow 2NaOH(aq) + H_2(g)$
(b) $2K(s) + Br_2(l) \rightarrow 2KBr(s)$
(c) $6Li(s) + N_2(g) \rightarrow 2Li_3N(s)$
(d) $2Na(s) + H_2(g) \rightarrow 2NaH(s)$
(e) $NaH(s) + H_2O(l) \rightarrow NaOH(aq) + H_2(g)$
3-6. (a) $4Li(s) + O_2(g) \rightarrow 2Li_2O(s)$
(b) $2Na(s) + O_2(g) \rightarrow Na_2O_2(s)$
(c) $K(s) + O_2(g) \rightarrow KO_2(s)$
(d) $Cs(s) + O_2(g) \rightarrow CsO_2(s)$
3-7. (a) $4KO_2(s) + 2H_2O(g) \rightarrow 3O_2(g) + 4KOH(s)$
(b) $2Na_2O_2(s) + 2CO_2(g) \rightarrow 2Na_2CO_3(s) + O_2(g)$
(c) $NaOH(s) + CO_2(g) \rightarrow NaHCO_3(s)$
(d) $NaNH_2(s) + H_2O(l) \rightarrow NaOH(aq) + NH_3(g)$
3-8. $2Na(l) \rightarrow 2Na^+ + 2e^-$, anode; $S(l) + 2e^- \rightarrow S^{2-}$, cathode;
$2Na(l) + S(l) \rightarrow Na_2S(l)$
3-9. Hydrogen is formed at a lower voltage than sodium. Even if Na^+ were reduced to Na(s), the Na(s) would react with water.
3-10. $2Na(amm) + 2NH_3(l) \xrightarrow{Fe_2O_3} 2NaNH_2(amm) + H_2(g)$
3-11. Potassium, rubidium, and cesium give superoxides.
3-12. The superoxide ion has an odd number of electrons. The ground-state electron configuration of O_2^{2-} is $(1\sigma)^2(1\sigma^*)^2(2\sigma)^2(2\sigma^*)^2(1\pi)^4(3\sigma)^2(1\pi^*)^4$. There are no unpaired electrons.
3-13. 506 g
3-14. 363 g
3-15. 122 L
3-16. 126 L
3-17. Li, 91.4 $J \cdot K^{-1} \cdot mol^{-1}$; Na, 85.0 $J \cdot K^{-1} \cdot mol^{-1}$;
K, 75.5 $J \cdot K^{-1} \cdot mol^{-1}$; Rb, 78.4 $J \cdot K^{-1} \cdot mol^{-1}$;
Cs, 71.1 $J \cdot K^{-1} \cdot mol^{-1}$

Chapter 4

4-1. (a) $Ca(s) + H_2(g) \xrightarrow{500°C} CaH_2(s)$
(b) $3Mg(s) + N_2(g) \xrightarrow{500°C} Mg_3N_2(s)$
(c) $Sr(s) + S(s) \xrightarrow{500°C} SrS(s)$
(d) $Ba(s) + O_2(g) \xrightarrow{500°C} BaO_2(s)$
4-2. (a) $Ca(s) + 2H_2O(l) \rightarrow Ca(OH)_2(aq) + H_2(g)$
(b) $Sr_3N_2(s) + 6H_2O(l) \rightarrow 3Sr(OH)_2(aq) + 2NH_3(g)$
(c) $CaC_2(s) + 2H_2O(l) \rightarrow Ca(OH)_2(s) + C_2H_2(g)$
(d) $Ca(s) + 2C(s) \xrightarrow{500°C} CaC_2(s)$
4-3. $Be(s) + HCl(aq) \rightarrow BeCl_2(aq) + H_2(g)$
(b) $Be(s) + 2NaOH(aq) + 2H_2O(l) \rightarrow Na_2Be(OH)_4(aq) + H_2(g)$
(c) $3Be(s) + N_2(g) \xrightarrow{500°C} Be_3N_2(s)$
(d) $2Be(s) + O_2(g) \xrightarrow{400°C} 2BeO(s)$
4-4. Hot magnesium reacts with both water and carbon dioxide.
4-5. $Be(OH)_2(s) + 2HCl(aq) \rightarrow BeCl_2(aq) + 2H_2O(l)$;
$Be(OH)_2(s) + 2NaOH(aq) \rightarrow Na_2Be(OH)_4(aq)$ [or $Na_2BeO_2(aq) + 2H_2O(l)$]
4-6. (a) $Mg(OH)_2(s) + 2HCl(aq) \rightarrow MgCl_2(aq) + 2H_2O(l)$
(b) 2.92 mg
4-7. $BeCl_2(l) \rightarrow Be(l) + Cl_2(g)$; $K_2BeF_4(l) \rightarrow 2KF(l) + Be(s) + F_2(g)$;
$BeF_2(s) + Mg(s) \rightarrow MgF_2(s) + Be(s)$
4-8. $2BaO(s) + O_2(g) \rightarrow 2BaO_2(s)$;
$BaO_2(s) + 2HCl(aq) \rightarrow BaCl_2(aq) + H_2O_2(aq)$
4-9. $MgCO_3(aq) + 2HCl(aq) \rightarrow MgCl_2(aq) + H_2O(l) + CO_2(g)$
4-10. $CaCO_3(s) + 2HNO_3(aq) \rightarrow Ca(NO_3)_2(aq) + H_2O(l) + CO_2(g)$
4-11. (a) Linear (b) Combine each *sp* orbital on the beryllium atom with a *p* orbital on a chlorine atom.
4-12. (a) Tetrahedral (b) Combine each sp^3 orbital on the beryllium atom with a *p* orbital or a fluorine atom.
4-13. $Mg(OH)_2$: pH = 10.2; $Ca(OH)_2$: pH = 12.7; $Sr(OH)_2$: pH = 13.2; $Ba(OH)_2$: pH = 13.8
4-14. The key reaction is $BaCO_3(s) + 2HCl(aq) \rightarrow BaCl_2(aq) + H_2O(l) + CO_2(g)$.

4-15. (a)
```
           H
           |
CH3CH2—C—CH3
           |
          OH
```
(b)
```
          CH3
           |
CH3CH2—C—OH
           |
          CH3
```

Chapter 5

5-1. (a) $2Al(s) + Mn_2O_3(s) \rightarrow Al_2O_3(s) + 2Mn(l)$
(b) $Al(C_2H_3O_2)_3(s) + 3H_2O(l) \rightarrow Al(OH)_3(s) + 3HC_2H_3O_2(aq)$
(c) $Al(NO_3)_3(aq) + 3NH_3(aq) \rightarrow Al(OH)_3(s) + 3NH_4NO_3(aq)$
(d) $Ga(OH)_3(s) + KOH(aq) \rightarrow KGa(OH)_4(aq)$
5-2. The acidities of the oxides generally decrease upon descending a group.
5-3. (a) $2Ga(s) + 6HCl(aq) \rightarrow 2GaCl_3(aq) + 3H_2(g)$
(b) $2Ga(s) + 2NaOH(aq) + 6H_2O(l) \rightarrow 2NaGa(OH)_4(aq) + 3H_2(g)$
5-4. $Al(OH)_3(s) + 3H^+(aq) \rightarrow Al^{3+}(aq) + 3H_2O(l)$;
$Al(OH)_3(s) + OH^-(aq) \rightarrow Al(OH)_4^-(aq)$.
For $Ga(OH)_3$ replace Al with Ga in these equations.
5-5. $B(OH)_3(aq) + H_2O(l) \rightarrow H^+(aq) + B(OH)_4^-(aq)$
5-6. $5B_2H_6(g) \xrightarrow{pyrolysis} B_{10}H_{14}(s) + 8H_2(g)$
5-7. $4BF_3(g) + 3NaBH_4(soln) \rightarrow 3NaBF_4(soln) + 2B_2H_6(soln)$
5-8. (a) 18 (b) 42
5-9. (a) Octahedral (b) tetrahedral (c) linear
5-10. (a) trigonal planar (b) linear (c) tetrahedral
5-11. tetrahedral; combine each sp^3 hybrid orbital on the boron atom with a *p* orbital on a fluorine atom

5-12. Use sp^3 orbitals on each boron atom and form three-center bonds as in B_2H_6 (Figure 5-4).
5-13. Use sp^3 orbitals on each boron atom and form three-center bonds as in B_2H_6 (Figure 5-4).
5-14. $Al_2(SO_4)_3(aq) + 3Na_2CO_3(aq) + 3H_2O(l) \rightarrow$
$2Al(OH)_3(s) + 3Na_2SO_4(aq) + 3CO_2(g)$
The flocculent precipitate of $Al(OH)_3$ carries down particulate matter as it settles.
5-15. $K = 0.044\ M^{-4} \cdot atm^{-1}$
5-16. It requires 230 kJ to bring the reaction products to the melting point of chromium and to melt the chromium. In either case, the chromium will be melted.
5-17. $Al^{3+}(aq)$: pH = 2.98; $Ga^{3+}(aq)$: pH = 1.83; $In^{3+}(aq)$: pH = 1.86; $Tl^{3+}(aq)$: pH = 1.25
5-18. 3.37

Chapter 6

6-1. (a) $2Al_2O_3(s) + 6C(s) \xrightarrow[\text{furnace}]{\text{electric}} Al_4C_3(s) + 3CO_2(g)$
(b) $Al_4C_3(s) + 12D_2O(l) \rightarrow 4Al(OD)_3(s) + 3CD_4(g)$
(c) $CaC_2(s) + 2D_2O(l) \rightarrow C_2D_2(g) + Ca(OD)_2(s)$
(d) $PbS(s) + O_2(g) \xrightarrow{\text{heat}} Pb(s) + SO_2(s)$
6-2. (a) $PbO(s) + CO(g) \rightarrow Pb(s) + CO_2(g)$
(b) $Si(s) + 4NaOH(aq) \rightarrow Na_4SiO_4(aq) + 2H_2(g)$
(c) $SiCl_4(g) + 4H_2O(l) \rightarrow SiO_2 \cdot 2H_2O(s) + 4HCl(aq)$
(d) $SiO_2(s) + 6HF(aq) \rightarrow H_2SiF_6(s) + 2H_2O(l)$
6-3. decrease
6-4. $Sn(OH)_2(s) + 2H^+(aq) \rightarrow Sn^{2+}(aq) + 2H_2O(l)$;
$Sn(OH)_2(s) + 2OH^-(aq) \rightarrow Sn(OH)_4^{2-}(aq)$
6-5. Diamond must have a smaller molar volume, and hence a greater density than graphite.
6-6. See Figures 6-2 and 6-3.
6-7. All the valence electrons in diamond are in localized covalent bonds and are tightly held. Graphite has many delocalized electrons (see Figure 6-3), which are able to carry an electric current.
6-8. Carbon dioxide in the air dissolves to form carbonic acid.
6-9. $^{\ominus}\ddot{N}{=}C{=}\ddot{N}^{\ominus}$; linear
6-10. $:N{\equiv}C{-}C{\equiv}N:$; linear
6-11. Heat either silicon or silicon dioxide with carbon in an electric furnace.
6-12. *n*-type: As, Sb; *p*-type: Ga, In
6-13. Carbon tetrahalides are saturated, meaning that the central carbon atom can form no more bonds. Silicon can use its *d* orbitals to add additional ligands to silicon tetrahalides, and hence silicon tetrahalides are much more reactive than carbon tetrahalides.
6-14. (a) Si: +4, C: −4 (b) +4 (c) +2 (d) +4
6-15. (a) +4 (b) 8/3 (or +4, +2, +2) (c) −4 (d) −1
6-16. (a)
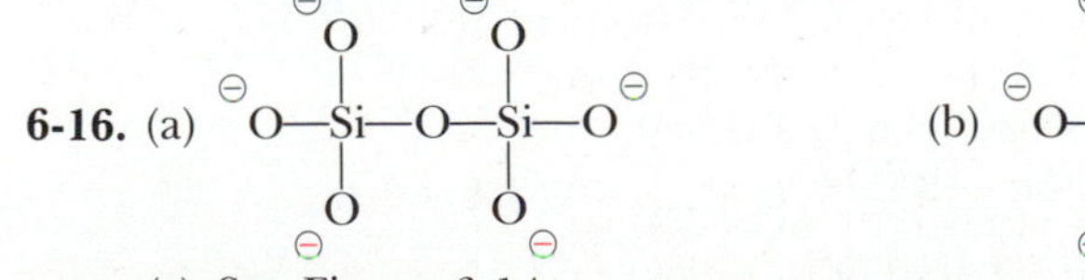
(b) [structure: O–Si–O–Si–O–Si–O chain, each Si bearing O⊖ above and below; terminal O⊖]
(c) See Figure 6-14.
6-17. $SiO_2(s) + 6HF(aq) \rightarrow H_2SiF_6(aq) + 2H_2O(l)$
6-18. (a) tetrahedral (b) bent (c) octahedral (d) trigonal pyramidal

6-19. SiH_4: similar to methane (see Figure 12-5 of the text); combine each sp^3 orbital on the silicon atom with an s orbital on a hydrogen atom. Si_2H_6: similar to ethane (see Figure 12-7 of the text).
6-20. tin: $SnO_2(s) + C(s) \rightarrow Sn(s) + CO_2(g)$;
lead: $2PbS(s) + 3O_2(g) \rightarrow 2PbO(s) + 2SO_2(g)$
or $PbO(s) + CO(g) \rightarrow Pb(s) + CO_2(g)$
6-21. See page 65.
6-22. If the temperature of the tin objects is allowed to fall below 13°C for prolonged periods, then the brittle allotrope, gray tin, is formed and the object disintegrates.
6-23. 86.6 kg
6-24. 78.8 kg

Chapter 7

7-1. (a) $P_4(s) + \underset{\text{excess}}{5O_2(g)} \rightarrow P_4O_{10}(s)$
(b) $P_4O_6(s) + 6H_2O(l) \rightarrow 4H_3PO_3(aq)$
(c) $P_4O_{10}(s) + 6H_2O(l) \rightarrow 4H_3PO_4(aq)$
(d) $2NaN_3(s) \xrightarrow{\text{heat}} 2Na(s) + 3N_2(g)$
7-2. (a) $N_2O_3(g) + H_2O(l) \rightarrow 2HNO_2(aq)$
(b) $N_2O_5(s) + H_2O(l) \rightarrow 2HNO_3(aq)$
(c) $NH_4NO_3(s) \xrightarrow{\text{heat}} N_2O(g) + 2H_2O(l)$
(d) $NH_4NO_2(s) \xrightarrow{\text{heat}} N_2(g) + 2H_2O(l)$
7-3. $C(s) + H_2O(g) \xrightarrow{\text{heat}} CO(g) + H_2(g)$ (water-gas reaction);
$N_2(air) + 3H_2(g) \xrightarrow[500°C]{300\text{ atm}} 2NH_3(g)$ (Haber process)
7-4. $Li_3N(s) + 3D_2O(l) \rightarrow 3LiOD(s) + ND_3(g)$
7-5. $N_2H_4(aq) + HNO_2(aq) \rightarrow HN_3(aq) + 2H_2O(l)$
7-6. $2NH_3(aq) + ClO^-(aq) \xrightarrow{OH^-(aq)} N_2H_4(aq) + H_2O(l) + Cl^-(aq)$
7-7. Solutions of nitric acid often have a brown-yellow color because the nitric acid slowly decomposes in the presence of light, producing NO_2.
7-8. Nitrogen can use only its $2s$ and $2p$ orbitals for bonding, and thus can form a maximum of four bonds.

7-9. (a) $O{=}N^{\oplus}(O^{\ominus}){-}N^{\oplus}(O^{\ominus}){=}O$ + resonance forms (b) $O{=}N{-}N^{\oplus}({=}O){-}O^{\ominus}$

(c) $^{\ominus}N{=}C{=}N^{\ominus}$ (d) $H{-}N(H){-}N(H){-}H$ (e) $^{\ominus}N{=}N^{\oplus}{=}N^{\ominus}$

7-10. (a) $H{-}N(H){-}O{-}H$ (b) $^{\ominus}O{-}N^{\oplus}({=}O){-}O^{\ominus}$ + resonance forms

(c) $^{\ominus}O{-}N{=}O$ (d) $^{\ominus}N{=}N^{\oplus}{=}O$ (e) $^{\ominus}O{-}N^{\oplus}{=}O$

7-11. $PCl_3(l) + 3D_2O(l) \rightarrow 3DCl(g) + D_3PO_3(aq)$ or
$PCl_5(s) + 4D_2O(l) \rightarrow 5DCl(g) + D_3PO_4(aq)$
7-12. $Ca_3P_2(s) + 6D_2O(l) \rightarrow 2PD_3(g) + 3Ca(OD)_2(s)$
7-13. $P_4O_{10}(s) + 6H_2SO_4(l) \rightarrow 4H_3PO_4(l) + 6SO_3(g)$;
$P_4O_{10}(s) + 12HNO_3(l) \rightarrow 4H_3PO_4(l) + 6N_2O_5(s)$
7-14. bent
7-15. (a) tetrahedral (b) tetrahedral (c) octahedral

7-16. $P_4O_6S_4$: The four sulfur atoms take the place of the four terminal oxygen atoms (see Figure 7-8b).
$P_4O_4S_6$: The six sulfur atoms take the place of the six nonterminal oxygen atoms (see Figure 7-8b).
7-17. See page 90.
7-18. See page 95.
7-19. (a) $SF_4(g) + 3H_2O(l) \rightarrow H_2SO_3(aq) + 4HF(g)$
(c) $XeF_6(s) + 3H_2O(l) \rightarrow XeO_3(aq) + 6HF(g)$
(d) $BrF_5(l) + 3H_2O(l) \rightarrow HBrO_3(aq) + 5HF(g)$
(e) $AsBr_3(s) + 3H_2O(l) \rightarrow H_3AsO_3(aq) + 3HBr(g)$
7-20. (a) trigonal pyramidal (b) tetrahedral (c) trigonal pyramidal (d) trigonal pyramidal (e) trigonal bipyramidal
7-21. As the central atom becomes larger, the tendency to form hybrid orbitals decreases, and so the bonding in the heavier hydrides can be described by combining a *p* orbital on the central atom with an *s* orbital on a hydrogen atom.
7-22. The mixture becomes increasingly colored with inceasing temperature (see p. 564 of the text).
7-23. (a) $PCl_3(l) + 3D_2O(l) \rightarrow 3DCl(g) + D_3PO_3(l)$; 2.67 mg
(b) 0.803 atm
7-24. (a) 1.65 (b) 1.49 (c) 1.27

Chapter 8

8-1. photosynthesis
8-2. $2KClO_3(s) \xrightarrow{MnO_2(s)} 2KCl(s) + 3O_2(g)$;
$2Na_2O_2(s) + 2H_2O(l) \rightarrow 4NaOH(aq) + O_2(g)$
8-3. $2C_2H_2(g) + 5O_2(g) \rightarrow 4CO_2(g) + 2H_2O(g)$
8-4. self-contained breathing apparatus
8-5. $PCl_3(l) + 3D_2O(l) \rightarrow 3DCl(g) + D_3PO_3(aq)$;
$Na_2O_2(s) + 2DCl(aq) \rightarrow 2NaCl(aq) + D_2O_2(aq)$
8-6. Prepare $H_2{}^{18}O_2$ by burning sodium in $^{18}O_2$ to make $Na_2{}^{18}O_2$ and then reacting this with HCl. If the oxygen evolved is $^{18}O_2$, then it must have come from the $H_2{}^{18}O_2$. The presence of $^{18}O_2$ can be determined by means of a mass spectrometer.
8-7. See Figure 8-1.
8-8. See Figure 8-6.
8-9. See Section 8-10.
8-10. Concentrated sulfuric acid is too strong an acid; it would destroy the root systems of the plants.
8-11. SF_6 is a relatively large and massive molecule, so SF_6 will effuse and diffuse more slowly through the wall of the ball than other gases (see Graham's law of effusion).

8-12. F—S—S—F   F—S(F)=S

8-13. O=S(=O)(—O—H)—O—S(=O)(—O—H)=O

8-14. H—O—S(=O)(=O)—O—O—H

8-15. $2S_2O_3^{2-}(aq) + I_3^-(aq) \rightarrow S_4O_6^{2-}(aq) + 3I^-(aq)$
8-16. $4Ag(s) + 2H_2S(g) + O_2(g) \rightarrow 2Ag_2S(s) + 2H_2O(l)$
8-17. $2H_2S(g) + 3O_2(g) \rightarrow 2SO_2(g) + 2H_2O(l)$;
$SO_2(g) + 2H_2S(g) \rightarrow 3S(s) + 2H_2O(l)$
8-18. See answer to Problem 7-21.
8-19. The bonding in SF_6 is saturated, meaning that the sulfur atom can bond to no more ligands. Small molecules or ions can bond to the sulfur atom in SF_4, however, providing a mechanistic pathway for reactions.
8-20. $K_p = 3.48 \times 10^{65}$ atm^{-1}. The production of SF_6 is more favored at low temperatures.
8-21. 0.611 M
8-22. 1.13×10^{-4} mol
8-23. 2.31×10^{-4} mol

Chapter 9

9-1. Fluorine is the strongest oxidizing agent and thus can remove electrons better than any other substance.
9-2. Fluorine is prepared by the electrolysis of a solution of HF in molten KF. Chlorine is prepared by electrolysis of molten NaCl. Bromine is prepared from certain brines by oxidation of $Br^-(aq)$ by chlorine. Iodine is prepared in a manner similar to bromine.
9-3. (a) $Ca(s) + Br_2(l) \rightarrow CaBr_2(s)$
(b) $Ti(s) + 2Cl_2(g) \rightarrow TiCl_4(l)$
(c) $2As(s) + \underset{\text{excess}}{3Cl_2(g)} \rightarrow 2AsCl_3(l)$
(d) $2Na(s) + I_2(s) \rightarrow 2NaI(s)$
9-4. (a) $2F_2(g) + H_2O(l) \rightarrow OF_2(g) + 2HF(g)$
(b) $2F_2(g) + Si(s) \rightarrow SiF_4(g)$
(c) $F_2(g) + 2Ag(s) \rightarrow 2AgF(s)$
(d) $F_2(g) + Mg(s) \rightarrow MgF_2(s)$
9-5. (a) $4NaCl(aq) + 2H_2SO_4(aq) + MnO_2(s) \rightarrow$
$2Na_2SO_4(aq) + MnCl_2(aq) + 2H_2O(l) + Cl_2(g)$
(b) $4NaIO_3(aq) + 10NaHSO_3(aq) \rightarrow$
$2I_2(s) + 7Na_2SO_4(aq) + 3H_2SO_4(aq) + 2H_2O(l)$
(c) $6Br_2(l) + 12NaOH(aq) \rightarrow 10NaBr(aq) + 2NaBrO_3(aq) + 6H_2O(l)$
9-6. $5Cl_2(g) + I_2(s) + 6H_2O(l) \rightarrow 10HCl(aq) + 2HIO_3(aq)$
9-7. $3I_2(s) + 10HNO_3(aq) \rightarrow 6HIO_3(aq) + 10NO(g) + 2H_2O(l)$
9-8. $HClO_4$, perchloric acid; $HClO_3$, chloric acid; $HClO_2$, chlorous acid; HClO, hypochlorous acid
9-9. (a) bromous acid (b) hypoiodous acid (c) perbromic acid (d) iodic acid
9-10. oxyacids: (a) nitrous acid (b) sulfurous acid (c) hypophosphorous acid (d) phosphorous acid (e) hyponitrous acid; salts: (a) potassium sulfite (b) calcium nitrite (c) potassium iodite (d) magnesium hypobromite
9-11. $I_2(s) + 5H_2O_2(aq) \rightarrow 2HIO_3(aq) + 4H_2O(l)$;
$HIO_3(aq) + KOH(aq) \rightarrow KIO_3(aq) + H_2O(l)$
9-12. The substances HCl(*aq*), HBr(*aq*), and KOH(*aq*) are strong electrolytes, so the species $K^+(aq)$, $Cl^-(aq)$, and $Br^-(aq)$ are spectator ions. The only reaction in each case is $H^+(aq) + OH^-(aq) \rightarrow H_2O(l)$; HF(*aq*) is a weak acid, and some energy goes into dissociating HF(*aq*) into $H^+(aq)$ and $F^-(aq)$.
9-13. (a) $IO_4^-(aq) + H^+(aq) + 2H_2O(l) \rightarrow H_5IO_6(aq)$
(b) $5H_5IO_6(aq) + 2Mn^{2+}(aq) \rightarrow 2MnO_4^-(aq) + 5IO_3^-(aq) + 11H^+(aq) + 7H_2O(l)$
9-14. $6I_2(s) + 12OH^-(aq) \rightarrow 10I^-(aq) + 2IO_3^-(aq) + 6H_2O(l)$

9-15. Water is oxidized to oxygen at a lower voltage than that required to oxidize $F^-(aq)$ to $F_2(g)$.
9-16. $3IF(g) \rightleftharpoons IF_3(g) + I_2(s)$
9-17. Fluorine, being so electronegative, produces a positive partial positive charge on the nitrogen atom. Thus, the lone pair is held so strongly in NF_3 that it is not able to act as a Lewis base.
9-18. 0
9-19. The structure of Cl_2O_7 is O—Cl(=O)(—O)—O—Cl(—O)(—O)—O (each Cl bonded to O above and O below). The bonding can be described in terms of sp^3 orbitals on all nine atoms.
9-20. (a) bent (b) seesaw (c) octahedral
9-21. (a) I: +5, F: −1 (b) +1 (c) +5 (d) Cl: +1, F: −1 (e) +5
9-22. [HF] = 0.0496 M; $[F^-]$ = 0.0335 M; $[HF_2^-]$ = 0.00846 M
9-23. 697 M^{-1}
9-24. 166 M^{-1}
9-25. 0.875 millimole
9-26. 0.0489 = 0.0476 + 0.0013
9-27. $K = 7.6 \times 10^{-10}\ M^3 \cdot atm^{-1}$
9-28. $K = 1.2 \times 10^{-8}\ M^2$

Chapter 10

10-1. Their lack of reactivity.
10-2. They were discovered as a result of an investigation of the source of a discrepancy between the density of pure nitrogen and that of air from which all the oxygen, water vapor, and carbon dioxide had been removed.
10-3. natural gas
10-4. They have similar chemical properties which differ from all the other groups of elements.
10-5. radioactive decay, producing α-particles
10-6. See Figure 10-1. The remaining $N_2(g)$ can be removed by reacting with lithium.
10-7. Their lack of reactivity implies unusually stable electron configurations.
10-8. Lithium reacts with nitrogen, but not with argon.
10-9. The ionization energies of O_2 and Xe are about the same.
10-10. (a) $6XeF_4(s) + 12H_2O(l) \rightarrow 2XeO_3(s) + 4Xe(g) + 3O_2(g) + 24HF(aq)$
(b) $Xe(g) + F_2(g) \rightarrow XeF_2(s) \xrightarrow{F_2(g)} XeF_4(s) \xrightarrow{F_2(g)} XeF_6(s)$
(c) $Kr(g) + F_2(g) \rightarrow KrF_2(s)$
(d) $XeF_6(s) + 3H_2O(l) \rightarrow XeO_3(s) + 6HF(g)$
10-11.

	He	Ne	Ar	Kr	Xe
$\Delta\bar{S}_{fus}/J \cdot K^{-1} \cdot mol^{-1}$	—	13.6	14.0	14.1	14.3
$\Delta\bar{S}_{vap}/J \cdot K^{-1} \cdot mol^{-1}$	19.0	64.9	74.7	75.4	76.5

The value of $\Delta\bar{S}_{vap}$ according to Trouton's rule is 85 $J \cdot K^{-1} \cdot mol^{-1}$.
10-12. The van der Waals constants *a* and *b* are related to the attraction between molecules and to their sizes, respectively. Both of these quantities increase with increasing atomic number.
10-13. The attraction between neutral molecules increases with the number of electrons (see Section 13-4 of the text).
10-14. See Table 10-4.
10-15. XeF_2: +2; XeF_4: +4; XeF_6: +6; $XeOF_4$: +6; XeO_2F_2: +6; XeO_3: +6
10-16. (a) linear (b) square planar (c) trigonal pyramidal (d) tetrahedral

Chapter 11

11-1. tungsten
11-2. iridium and osmium
11-3. iron
11-4. scandium and titanium
11-5. $TiO_2(s) + C(s) + 2Cl_2(g) \xrightarrow{1000°C} CO_2(g) + TiCl_4(g)$;
$TiCl_4(g) + 2Mg(s) \rightarrow 2MgCl_2(s) + Ti(s)$
11-6. Sc: +3; Ti: +4; V: +5; Cr: +6; Mn: +7
11-7. vanadium pentoxide
11-8. 30.2 g
11-9. $2Fe(s) + Cr_2O_7^{2-}(aq) + H_2O(l) \rightarrow Cr_2O_3(s) + Fe_2O_3(s) + 2OH^-(aq)$
11-10. $Zn(s) + 2Cr^{3+}(aq) \rightarrow 2Cr^{2+}(aq) + Zn^{2+}(aq)$;
$4Cr^{2+}(aq) + O_2(aq) + 4H^+(aq) \rightarrow 4Cr^{3+}(aq) + 2H_2O(l)$
11-11. Potassium permanganate slowly oxidizes the water according to: $4MnO_4^-(aq) + 2H_2O(l) \rightarrow 4MnO_2(s) + 3O_2(g) + 4OH^-(aq)$.
This reaction is much more rapid in the presence of light (photochemical reaction).
11-12. 3.47 g
11-13. (a) tetrahedral (b) trigonal bipyramidal
(c) tetrahedral (d) tetrahedral
11-14. See Figure 11-5.
11-15. See Figure 11-6.
11-16. 3500 metric tons of iron; 9670 metric tons of slag
11-17. Aluminum forms an impervious coating of Al_2O_3.
11-18. See Figure 11-8.
11-19. Cobalt(III) oxidizes water.
11-20. selenium and tellurium
11-21. 109 metric tons
11-22. The $NO(g)$ that is evolved reacts immediately with the oxygen in the air to produce $NO_2(g)$, which is red-brown.
11-23. 58.3%
11-24. See pages 158 and 159.

INDEX